An Encyclopedia Of Concepts In MRI Safety

EW Wooten

A Reference For The MRSO, MRMD, and MRSE

REKM Publishing
Second Edition

Made in the United States of America

Library of Congress-in-Publication Data

Wooten, Edward Wrenn

p. cm.

An Encyclopedia Of Concepts In MRI Safety:
A Reference For The MRSO, MRMD, and MRSE

Includes bibliographical references.

ISBN 978-0-578-91050-5

1. Magnetic resonance imaging. 2. Magnetic resonance imaging- safety measures.
3. Magnetic resonance imaging- complications. 4. Magnetic resonance imaging- physics.

05142021

This book was designed to provide a general reference for those working in and around the MR environment and/or providing patient care such as radiologists, MRI technologists, department/facility administrators, MRI physicists, MRI researchers, and others. The information in this book is only for general informational purposes, and not intended to provide medical or technical recommendations, opinions, or advice on any specific instances, situations, or facts for individuals or organizations, including but not limited to patients, technologists, hospitals, clinics, and medical pro-viders including but not limited to physicians. The general information provided should not be considered specific to any individual or individual situation, whether or not a specific patient is referenced by any individual, group, or entity in pursuit of information. The contents of this work should not be regarded as recommending or promoting any specific method, diagnostic strategy, treatment strategy, or management strategy by any physician, technologist, other healthcare or MR practitioners, or any other individual or group in any particular situation. The author/publisher of this work disclaims any liability for the actions of any individual, group, or entity acting independently or on behalf of any individual, any group, or any organization utilizing the information in this book.

The author/publisher makes no representations or warranties with respect to the accuracy or completeness of the contents of this work and specifically disclaims all warranties, including without limitation any expressed or implied warranties of appropriateness of the information in this book for a particular purpose. Although effort has been made to ensure accuracy, the author/publisher assumes no responsibilities for errors or omissions which may be related to conceptual, technical or other inaccuracies, or typographical or grammatical errors, or any other error.

The cover design is based on an algorithm by Henrik Schumacher, used with permission.

Comments may be sent to mrisafetyencyclopedia@gmail.com.

To Kimberly, Robert, Elizabeth, Katherine, and Mary

About the Author

E. Wrenn Wooten, MD, DPhil, MRSO/MRMD/MRSE (MRSC™) attended the University of Oxford as a Rhodes Scholar where he received his doctorate studying the development and application of nuclear magnetic resonance spectroscopy as well as quantum mechanical and semiclassical energy calculations to biophysical systems. He did postdoctoral fellowships in MRI physics at the University of Arkansas for Medical Sciences and in chemical physics at the University of California at Berkeley. He received his MD from Vanderbilt University, and did his residency in diagnostic radiology at the Mallinckrodt Institute of Radiology at Washington University and a clinical MRI fellowship at Baylor University Medical Center. He is board certified in diagnostic radiology by the American Board of Radiology and magnetic resonance safety certified (MRSC™) by the American Board of Magnetic Resonance Safety with certifications as a Magnetic Resonance Safety Officer, Magnetic Resonance Medical Director, and Magnetic Resonance Safety Expert.

Preface to the Second Edition

With the update of the ACR Manual on MR Safety in 2020, some of the information in the first edition of An Encyclopedia Of Concepts In MRI Safety has become obsolete and has required updating. The 2018 ACR Manual on Contrast Media has also been replaced by a 2020 version. In addition, new guidelines from ICNIRP and ASTM are also incorporated. Based on questions from some readers, additional entries have been made and new references are cited. Since the writing of the first edition I acknowledge helpful discussions with and/or suggestions from Norman Cheng, Christopher Collins, Ryan Draeger, Cihat Eldeniz, Tobias Gilk, David Hoult, Gregory Hurst, Bernd Ittermann, Jianming Jiang, Emmanuel Kanal, John Kirsch, Kirk McDonald, Grey Naughton, and Daniel Sodickson.

Preface to the First Edition

The issue of MRI safety is not insignificant, as the rate of MRI related accidents is increasing faster than the rate of MRI examinations, suggesting a significant MRI safety problem (Hardy (2015), Delfino et al. (2019)). To help combat this, The American Board of Magnetic Resonance Safety (ABMRS) was established in 2014 to help train and certify individuals in order to address both adverse outcomes related to MRI scanners as well as inappropriate denial of care when a patient with an implant or foreign body could be safely scanned but is not. The ABMRS offers three certifications: the MR Safety Officer (MRSO), MR Medical Director (MRMD), and the MR Safety Expert (MRSE). All of these are MR Safety Certified (MRSC™). Most MRSOs are MRI technologists, all MRMDs are required to be medical doctors, and most MRSEs are MRI physicists, although there are no specific requirements for the MRSO and MRSE certifications other than to pass the ABMRS certification examination.

While there are numerous articles and books and some courses which contain information relevant to MRI safety and ABMRS certification, there seemed to be a need for a quick reference where basic information was collated and readily accessible both as a reference in preparing for the ABMRS examinations and as a reference for practitioners. This work is designed for that purpose and started as a set of study notes for the MRMD and MRSE examinations of the ABMRS.

The book can be used as a reference to look up a term, where the entry provides a brief definition and how it is relevant to MRI safety. In addition, a more quantitative discussion may follow in a given entry. In some cases example problems are also provided. Relevant literature citations are given. As such, the reference can also be read as a sort of textbook on MRI safety. Since the book is written with the MRSO, MRMD, and MRSE all in mind, it can be thought of as three books in parallel. Under the various entries, the initial information provided is felt to be broadly relevant to all three certifications. Information felt to be particularly relevant for the MRMD or the MRSE is then provided afterward, labeled appropriately, and can be skipped if felt less applicable to a particular practitioner. It is assumed that all readers are generally familiar with the principles and vocabulary of MRI. It is also assumed there is some familiarity with general medical principles.

For the MRMD, knowledge at the level of a physician is assumed. For the MRSE, familiarity with the principles of electromagnetics, vector calculus, linear algebra, and differential equations is assumed. The physics discussions and derivation are kept to a minimum, except where it was felt that a more extensive discussion was warranted in the cases of where the issues are more complex, or the information is difficult to find in detail in the extant literature, for example the near field, the magnetic displacement force, and the antenna effect.

I am fortunate to have had multiple discussions about magnetic resonance safety and the related physics with numerous experts in person, by telephone, and/or by email, who have been kind enough to patiently answer questions over the years. These individuals include Max Amurao, Renzo Bazzo, Eleana Bell, Norman Cheng, Michael Coey, Christopher Collins, Jana Delfino, Ivan Dimitrov, William Faulkner, David Fenyes, Peter Gettins, Tobias Gilk, Karl Glastad, David Griffiths, Steve Homans, David Hoult, Alexej Jershow, Brian Johnson, Emmanuel Kanal, John Kirsch, Richard Komoroski, Kirk McDonald, Donald McRobbie, John Nyenhuis, Johann Overweg, Larry Panych, Alexander Pines, Sunder Rajan, Matt Rederer, John Schenck, Cass Sackett, Francisco Sanchez, Jason Stafford, Terry Woods, Dmitriy Yablonskiy, and Andrew Zangwill.

I want to thank all of the MRI technologists who keep us all safe every day.

I especially want to thank my lovely, wonderful wife, Kimberly, who encouraged me to start and finish this project. I want to thank my four children for all of their support as well. I could not have done it without them.

Introduction

An MRI scanner is arguably one of the most dangerous pieces of equipment in a hospital or other imaging facility, not just in terms of the numerous ways that it can cause personal and property damage, but in the stealthy ways dangerous situations can sneak up until too late.

The safety issues involved in MRI may be roughly classified into the following mechanisms:

1. Force
2. Current
3. Pressure
4. Thermal
5. Chemical
6. Physical Confinement

These are somewhat arbitrary categories as there is some interplay between the various mechanisms.

In terms of forces the magnet may exert on objects there is an attractive displacement force, a rotational torque force, and a Lorentz force (sometimes referred to as a Lenz force). Currents may be induced through time varying magnetic fields or time varying electric fields with associated heating or implant damage/malfunction. Pressure may manifest itself as sound pressure level in the setting of acoustic noise or increased room pressure in the setting of a quench breach. Thermal effects may relate to heating of patients or object related heating during a scan or from accidents involving cryogens. Chemical effects may refer to complications from medication administration including gadolinium contrast agents or from the effect of breathing cryogens. Physical confinement relates to room entrapment in a quench breach. In addition, physical confinement may also refer to some patients experiencing feelings of claustrophobia within the physical confines of the scanner bore during a scan.

There is a natural tendency to try to apply the same sort of reasoning to MRI as would be applied to other aspects of medicine and to daily life in general. But, suppose that doubling the dose of a drug routinely produced sixteen times the effect, or pushing a drug in twice as fast quadrupled how fast it went in, or moving a scalpel over four centimeters produced a resulting incision which was only two centimeters long? That would seem strange in our relatively linear experience, but yet this is the counterintuitive world in which the MRI practitioner lives.

There are numerous examples of these types of nonlinear effects related to MRI safety, for example the attractive force with distance. An object could have minimal tug at one point near the scanner and then be uncontrollably tugged without warning only a few centimeters closer to the magnet. This is because of the nonintuitive fact that the attractive force depends on the highly nonlinear shape of the magnetic field around the magnet. In addition, the attractive force at the center of the bore is zero where the field is greatest, but maximal near the entrance of the bore. As another example, to the uninitiated the intuition is that a 180-degree pulse delivers two times the SAR of a 90-degree pulse, while the SAR of a 180-degree pulse is actually four times that of a 90-degree pulse. Another nonintuitive example is the well described experience of patients receiving significant burns from a scanner, but the patient was asymptomatic during the scan itself, when common sense would dictate that the patient should be able to warn the technologist of discomfort before significant damage is done. It is a source of confusion to many that a scan could potentially be safe at 3 T but dangerous at 1.5 T.

Nephrogenic systemic fibrosis is a disease which did not exist prior to the use of gadolinium contrast agents and is still incompletely understood, as are the effects of long term gadolinium retention in the body. Even this is medically atypical as a potentially life threatening complication may occur months or even years after exposure to only a single small dose of contrast.

In short, MRI and MRI safety are different from many other aspects of medicine. Understanding the relevant issues is crucial to keeping individuals and property safe from harm.

Mathematical Conventions

A right handed coordinate system is used. Vectors and tensors appear in bold type, for example the magnetic flux density vector field $\mathbf{B}_0$ and demagnetizing tensor $\mathbf{N}$. The corresponding scalar vector magnitudes appear as non-bolded italic text, for example the magnitude of the magnetic flux density, B_0. An exception to this is where confusion may arise, for example the magnitude of the gradient of the magnetic flux density, which is written $|\nabla B_0|$. The common convention for spherical coordinates in physics is used that θ is the angle between the z-axis and r while ϕ lies in the xy-plane. Note that for vector interactions the name of the interaction is sometimes used to represent the vector, vector field, or scalar magnitude. Consequently the magnetic flux density may refer to $\mathbf{B}_0(\mathbf{r})$, a more generic $\mathbf{B}_0$, and B_0. In addition, vector fields such as the magnetic flux density $\mathbf{B}(\mathbf{r})$ are explicitly functions of position $\mathbf{r}$, and sometimes this notation is explicitly used; however, most of the time the position dependence is understood when a notation such as $\mathbf{B}$ is used. Words such as "torque" may be used to refer to the vector or its magnitude. The common colloquialism of referring to $\mathbf{B}$ as the "magnetic field" is also employed at times. In addition, SAR and its time average $\overline{\text{SAR}}$ are often used interchangeably.

A

A. See Magnetic Vector Potential.

A-Weighting. A frequency dependent correction applied to sound pressure measurements made with a microphone to correct for the frequency dependent perception of the human ear (McJury (2014)). This is relevant in the setting of acoustic noise generated by an MRI scanner, for example the FDA and IEC limit for MRI acoustic noise is a root mean squared sound pressure level of 99 dB(A) with hearing protection in place, where the A denotes an A-weighted value (International Electrotechnical Commission (2010), FDA (2014), ACR Committee on MR Safety (2020)).

(*MRSE*) Example: Using a correction curve such as IEC 61672-1 2003 (International Electrotechnical Commission (2013)), there is -20 dB attenuation at 100 Hz frequency based on a measured calibration curve (Figure A.1). If 90 dB is measured, then the correction factor at 100 Hz is 20 dB and the A-weighting would be 90 dB $-$ 20 dB $=$ 70 dB(A). The dB(A) indicates the value has been A-weighted.

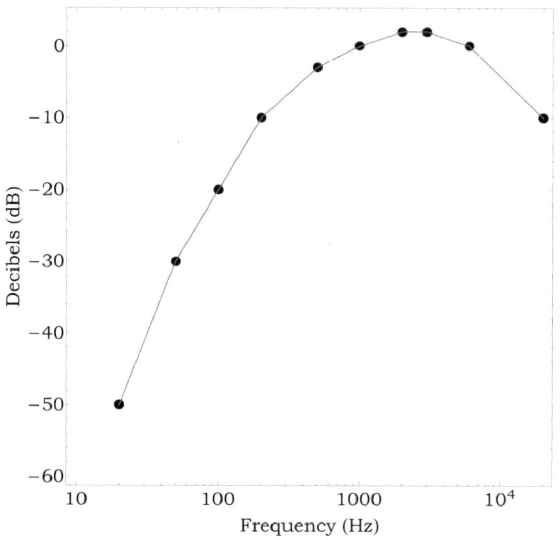

Figure A.1: A-weighting curve.

Abandoned Leads. Implants such as pacemakers or neural stimulator devices may have the battery pack removed, but the leads remain in place. Some studies have suggested these abandoned leads may represent a greater hazard for induced currents and heating during MRI compared to the complete device (Langman et al. (2011)). If the leads are capped off they may heat more than uncapped leads (Mattei et al. (2015)). Other studies have shown abandoned leads may be generally safe to scan (Horwood et al. (2017), Padmanabhan et al. (2018)). The Heart Rhythm Society 2017 Consensus Guidelines state that there are insufficient data to comment regarding the safety of MRI on patients with abandoned, epicardial, or fractured leads, although partially removed postsurgical temporary epicardial leads are not considered to be abandoned leads (Indik et al. (2017)). Partially removed postsurgical temporary epicardial leads do not appear represent a significant risk (Shellock (2020)).

ABMRS. See American Board of Magnetic Resonance Safety.

Acceleration Due to Gravity (g). The constant acceleration due to the Earth's gravitational field. The value is $g = 9.8\ m/s^2$. (Halliday et al. (2014)).

Accreditation In MRI. MRI facilities may be accredited by the American College of Radiology (ACR), Intersocietal Accreditation Commission, RadSite, and The Joint Commission (Clausen et al. (2018)). Accreditation from one of these organizations is required to qualify for Medicare reimbursement in accordance with the Medicare Improvements for Patients and Providers Act of 2008 (MIPPA).

Acoustic Noise. Acoustic noise is generated by an MRI scanner through the Lorentz force. When current flows through the time varying gradient coils, the coils move as a consequence of the Lorentz force. The motions are in the auditory range and can be loud enough to cause hearing loss (Delfino et al. (2019)). FDA acoustic noise limits are 140 dB peak unweighted and 99 dB(A) with hearing protection (McJury (2014), FDA (2014)). All patients and volunteers should be offered and encouraged to use hearing protection, with instruction for proper use and verification of fit and function prior to scanning. Provision of hearing protection similarly applies to any accompanying staff or companions remaining in Zone IV during scanning (ACR Committee on MR Safety (2020)).

ACR. See American College of Radiology.

ACR 2020 Manual on MR Safety. ACR publication which provides guidance on MRI safety (ACR Committee on MR Safety (2020)).

ACR 2015 MRI Quality Control Manual. Publication from the ACR addressing quality control in MRI (ACR (2015)). There are three sections including a Radiologist's section, an MRI Technologist's section, and an MRI Physicist/MRI Scientist's section. Assessment of MRI safety is included.

(*MRSE*) The most pertinent portion for MRI safety is Section F of the physicist's section regarding the MR Safety Program Assessment (ACR (2015)). Factors relevant to MRI hazards may be divided into facility design, operational, and clinical categories. Facility design relates to the physical design of the MRI suite, signage, and control of access to the various zones. Operational issues relate to the screening of people and objects. Finally, the clinical category refers to the medical management of implants and devices with respect to their suitability for scanning.

At least once a year the medical physicist must review the MRI site's written policies on MRI safety and ensure that the policies are accessible to facility staff. The following topics must be addressed in the policies: designated MR medical director, site access restrictions and MR zones, documented yearly MR safety training for all personnel, screening of patients and personnel, pediatric patients, magnet quenches, cryogen safety, acoustic noise, pregnant patients and staff, contrast media safety, sedation for examinations, thermal burns, medical emergency procedures, device and object screening, designation of MR status (Safe/Conditional/Unsafe), reporting of safety and adverse incidents, patient communication, and infection and medical waste control.

ACR 2020 Manual on Contrast Media. Publication from the ACR covering all aspects of contrast media (ACR Committee on Drugs and Contrast Media (2020)). For information on specific agents see the sections on Gadolinium.

Active Implant. Medical implant which uses electrical power (Nyenhuis (2014)). These can cause heating, be damaged, or malfunction during MRI (ACR Committee on MR Safety (2020)). Example: Pacemaker. If an active device is inadvertently to the fields associated with the MRI scanner, the physician responsible for the maintenance of the device should be contacted prior to the patient's discharge from the MRI suite. Significant injuries have resulted from even partial exposures, and adequate functionality should be verified for critical devices (ACR Committee on MR Safety (2020)).

Active Shielding. Coils placed around the main magnet coils which create a magnetic field opposite to the main magnetic field $\mathbf{B}_0$ causing the main magnetic field to decrease rapidly outside the scanner bore thereby decreasing the diameter of the fringe field (Junk (2014)). This allows for smaller scanner rooms, but leads to increased maximum spatial gradients and hence increased risk for missile effect accidents (McRobbie (2020)).

Acute Kidney Injury (AKI). Acute renal dysfunction within 48 hours of an inciting event. AKI alone is a risk factor for nephrogenic systemic fibrosis (NSF) (ACR Committee on Drugs and Contrast Media (2020)).

(*MRMD*) Acute kidney injury is diagnosed in the presence of the following within 48 hours of a nephrotoxic event (ACR Committee on Drugs and Contrast Media (2020)):

Absolute serum creatinine increase of $\geq 0.3\,mg/d\text{L}$.
Percentage increase in serum creatinine above baseline $\geq 50\%$.
Urine output reduced to $\leq 0.5\,m\text{L}/kg/h$ for at least 6 hours.

Administration. The 2020 ACR Guidelines (ACR Committee on MR Safety (2020)) as well as the 2018 ASRT Guidelines (Clausen et al. (2018)) assert that facility administrations have a duty to support and implement MR safe-practice guidelines established by the MRMD. Getting administrative support is often the most challenging aspect of developing, maintaining, and adhering to an MR safety program.

Adverse Event Reporting. All MR safety incidents or "near misses" are to be reported to the MR medical director in a timely manner, for example within 24 hours or one business day. The FDA also states that these incidents should be reported to them via their Medwatch program. This is a position supported by the ACR (ACR Committee on MR Safety (2020)). Local centers may have specific policies on reporting, however. A database of adverse events is available in the FDA MAUDE database.

AKI. See Acute Kidney Injury.

Allergic Reaction. A response to contrast which may be related to or clinically similar to an immune mediated hypersensitivity reaction. Allergic reaction symptoms to gadolinium may include erythema, urticaria (hives), nasal congestion, sneezing, itching, bronchospasm, edema (subcutaneous and/or laryngeal), hypotension, and anaphylactic shock (ACR Committee on Drugs and Contrast Media (2020)). See Gadolinium Contrast Reactions.

American Board of Magnetic Resonance Safety (ABMRS). Independent nonprofit organization which works to ensure the safety of all individuals in the MR environment through board certification in MR safety. After passing a board examination the individual becomes MR Safety Certified (MRSC™). Three levels of certification are available: Magnetic Resonance Safety Officer (MRSO), Magnetic Resonance Medical Director (MRMD), and Magnetic Resonance Safety Expert(MRSE) (Hardy (2015)).

American College of Radiology (ACR). A professional medical society representing physicians and medical physicists. The ACR provides accreditation of sites in the different imaging modalities including MRI. The ACR also has published the 2020 Manual on MR Safety, the 2015 MRI Quality Control Manual, and the 2020 Manual on Contrast Media.

American Society for Testing and Materials International (ASTM). (*MRSE*) An international body which develops voluntary technical standards for testing and classifying materials. Relevant to MRI safety there are standards for implants with respect to measuring and reporting the magnetic displacement force (F2052-15), the magnetic torque (F2213-17), rf heating of passive implants (F2182-19e2), and artifacts due to implants (F2119-07). In addition, there are standards for labeling objects with respect to being MR Safe, MR Unsafe, or MR Conditional (F2503-20) (ASTM F2052-15 (2015), ASTM F2119-07 (2013), ASTM F2182-19e2 (2019), ASTM F2213-17 (2017), ASTM F2503-20 (2020)).

American Society of Radiologic Technologists (ASRT). Organization which seeks to advance the medical imaging and radiation therapy professions and enhance safety and quality of patient care. This organization has published the Radiologic Technologist Best Practices for MR Safety (Clausen et al. (2018)).

Ampere (A). SI unit of current. 1 A$=$ 1 C/s (Halliday et al. (2014)).

Ampere's Law. One of Maxwell's Equations stating that a circulating magnetic field is associated with an electric current and/or a time varying electric field.

(*MRSE*). Ampere's Law in differential and integral form in a vacuum is (Zangwill (2013))

$$\nabla \times \mathbf{B} = \varepsilon_0 \mu_0 \frac{\partial \mathbf{E}}{\partial t} + \mu_0 \mathbf{J} \qquad \text{(A.1)}$$

and

$$\oint_C \mathbf{B} \cdot d\mathbf{l} = \mu_0 \int_S \mathbf{J} \cdot d\mathbf{A} + \varepsilon_0 \mu_0 \frac{\partial}{\partial t} \int_S \mathbf{E} \cdot d\mathbf{A} \quad \text{(A.2)}$$

where **B** is the magnitude of the magnetic flux density, **E** is the electric field, **J** is the current density, ε_0 is the electric permittivity of free space, and μ_0 is the magnetic permeability of free space.

Example: Find the magnetic field at a radial distance $r = 5\,mm$ from a straight thin wire carrying a current of $I = 1$ A along the wire. $\mu_0 = 1.257 \times 10^{-6}\ H/m$. The time dependent second term on the right hand side of Eq. A.2 is zero. Using the definition of current density

$$I = \int_S \mathbf{J} \cdot d\mathbf{A} \quad \text{(A.3)}$$

then

$$\oint_C \mathbf{B} \cdot d\mathbf{l} = \oint_C B dl \cos\theta = \mu_0 I \quad \text{(A.4)}$$

Since **B** and $d\mathbf{l}$ are in the same direction

$$Bl = \mu_0 I \quad \text{(A.5)}$$

For a circumference of radius $r = 5\,mm$ from the wire

$$B = \frac{\mu_0 I}{l} = \frac{\mu_0 I}{2\pi r} = \frac{(1.257 \times 10^{-6})(1\text{A})}{2\pi(5 \times 10^{-3}\,m)} = 4 \times 10^{-5}\,\text{T} \quad \text{(A.6)}$$

This provides the magnitude of the magnetic field around a straight wire. The direction of the field curling around the wire is determined from the right hand rule. Induced magnetic fields as a consequence of Ampere's Law can produce a reduction in the net field when combined with the $\mathbf{B}_1$ field and thus a decrease in SAR relative to a simple prediction from Faraday's Law (Panych and Madore (2018)).

Anesthesia. For various reasons patients may be under sedation including general anesthesia during an MRI examination. Guidelines for sedation have been developed by the American Academy of Pediatrics (AAP), American Society of Anesthesiologists (ASA), and The Joint Commission (TJC). These require preprocedure medical assessment, appropriate fasting, trained and credentialed sedation providers, monitoring with appropriate MR Conditional equipment, physical observation during the examination, resuscitation equipment, procedural medical charting, procedure for recovery and discharge criteria, and mechanism to track complications (Reinking (2014), ACR Committee on MR Safety (2020)).

Aneurysm Clip. A metallic clamp across a ballooning arterial structure in the brain. Motion of aneurysm clips due to forces from the magnet have led to bleeding and death. For this reason, particularly close attention to safety is necessary for these devices, and the ACR Manual on MR Safety has a separate section devoted to this (ACR Committee on MR Safety (2020)).

Prior plain film, CT, or MRI can clarify whether

there is an aneurysm clip. If prior imaging is not available, plain film or CT may be obtained to assess for the presence of a clip. It is necessary to obtain information on the manufacturer, model, and type of aneurysm clip in the patient to determine the safety status of the clip prior to imaging.

Note that all documentation regarding the clip must be be in writing and signed by a physician (non-physician signatures are not acceptable).

Aneurysm clips made after 1995 with current appropriate safety labeling may undergo MRI. Aneurysm clips made prior to 1995 require pretesting per the ASTM. Clips manufactured prior to 1995 require pretesting in accordance with ASTM methodology. Note that implantation date, in and of itself, is not sufficient to make a determination regarding safety for MR scanning without additional testing.

If there is a prior brain MRI available for review, a Level 2 radiologist may make an assessment regarding the magnetic properties based on susceptibility artifact.

If information is not available and for emergent clinical reasons a scan is contemplated, written consent with death listed as a potential complication must be obtained prior to the scan after conducting a suitable risk/benefit analysis (ACR Committee on MR Safety (2020)).

Angular Frequency (ω). (*MRSE*) The angular frequency in radians per second is related to the frequency in Hz by (Brown et al. (2014), Halliday et al. (2014))

$$\omega = 2\pi\nu \qquad \text{(A.7)}$$

ω is the angular frequency in radians/second and ν is the frequency in cycles/second (Hz).

Example: What is the angular frequency for a frequency of 60 MHz (60×10^6 Hz)?

$$\omega = 2\pi(60 \times 10^6\, Hz) = 3.77 \times 10^8 \text{ rad}/s. \qquad \text{(A.8)}$$

Anisotropy. A property which has different values in different directions exhibits anisotropy. For example, an ellipsoid has different length values in different directions. Properties such as magnetic susceptibility may be anisotropic (Cullity and Graham (2011)).

Antenna Effect. A conducting implant or potentially other foreign object can act as an antenna for the exciting rf field with associated heating of the adjacent tissue. The exciting rf field can create a standing wave current and voltage along the wire with voltage antinodes at the ends causing substantial heating. Nominally this occurs when the length of the object (such as an electrical lead) is half the wavelength λ of the rf in the surrounding medium such as tissue. This length is referred to as the resonant length. For 1.5 T this length is around 26 cm, and for 3 T around 13 cm in tissue. Note, however, that because of multiple uncertain factors potentially any length of material can have heating (ACR Committee on MR Safety (2020)).

The resonant length l_r in centimeters can therefore be estimated from B_0 in Tesla as

$$l_r \approx \frac{1.5}{B_0} 26 = \frac{39}{B_0} \tag{A.9}$$

(*MRSE*). The antenna effect is complex, and exact analytical solutions are elusive; however, a number of semiempirical and numerical treatments of the problem have been given (Panych and Madore (2018)). One of these is the transmission line model (Konings et al. (2000), Pictet et al. (2002), Nyenhuis (2014), Nyenhuis (2018)). A very simplified model is provided below.

Suppose that the electric field vector of an electromagnetic field from the transmit coil is oriented along the surface of a linear conductor such as a pacemaker lead, which is assumed to be a perfect conductor. The applied rf field exerts a force on the charges on the surface of the lead, which causes the charges to move, thereby creating a current. This current oscillates with the oscillating applied rf field. Associated with the current is an oscillating voltage. In turn, a second electric field is created which can cause heating of the adjacent tissues (Panych and Madore (2018), Nyenhuis (2018), Golestanirad et al. (2019)).

The lead can be modeled as a transmission line to see how the lead termination and length affect the electrical properties of the lead. As shown in Figure A.2, the lead is modeled as a two parallel wire transmission line of length l, extending from -l to 0. An open circuit configuration is assumed, and therefore the current and voltage waves are completely reflected, and standing waves of current and voltage are created.

The voltage and current on a lossless transmission line along z can be described as forward and backward traveling waves as follows (Hayt and Buck (2012), Nyenhuis (2018))

$$V(z,t) = V^{+}\cos(\omega t - kz) + V^{-}\cos(\omega t + kz) \tag{A.10}$$

and

$$I(z,t) = \frac{V^{+}}{Z_0}\cos(\omega t - kz) - \frac{V^{-}}{Z_0}\cos(\omega t + kz) \tag{A.11}$$

where $k = 2\pi/\lambda$, $V(z,t)$ is the voltage on the line, $I(z,t)$ is the current on the line, and V^{+} and V^{-} are the amplitudes of the forward and backward voltage waves, respectively. Z_0 is the characteristic impedance of the line. Since an open circuit is assumed, the forward and backward amplitudes are the same, Eqs A.10 and A.11 can be written in the commonly used phasor form (Hayt and Buck (2012), Nyenhuis (2018)) and converted into trigonometric form:

$$\widetilde{V}(z) = V^{+}\left(e^{-ikz} + e^{ikz}\right) = 2V^{+}\cos(kz) \tag{A.12}$$

and

$$\widetilde{I}(z) = \frac{V^+}{Z_0}\left(e^{-ikz} - e^{ikz}\right) = \frac{2iV^+}{Z_0}\sin(kz) \quad \text{(A.13)}$$

Note that at the ends of the line if the line length l is such that for $z = l = \lambda/2$ the voltage will have a resonant maximum amplitude as a function of length

$$\begin{aligned} |\widetilde{V}(0)| = |\widetilde{V}(l)| &= |2V^+\cos(kl)| \\ &= |2V^+\cos(\frac{2\pi}{\lambda}\frac{\lambda}{2})| = 2V^+ \end{aligned} \quad \text{(A.14)}$$

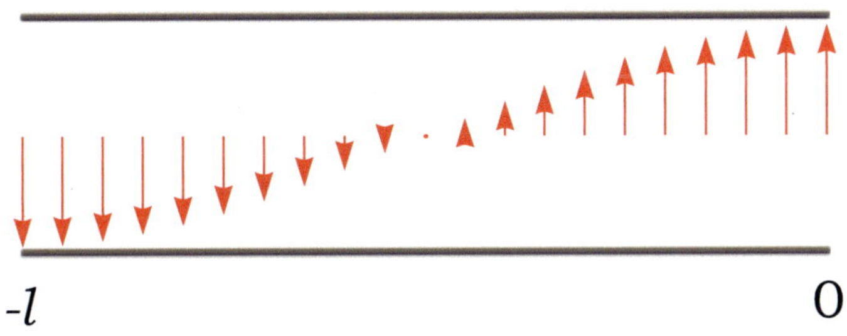

Figure A.2: Electric field along an open circuit parallel wire transmission line of length $l = \lambda/2$.

The voltage and current are out of phase by 90° in Eqs. A.12 and A.13. At the ends the voltage is maximal at $2\widetilde{V}^+$, while the current is zero. If the length of the lead l is $\lambda/2$ there is a maximum amplitude standing wave created. As the voltage is at a maximum at the electrode tip, the induced electric field will also be at a maximum in accordance with $E = V/l$, where l is the distance between the conductors (Steer (2019)). This is shown in Figure A.2. The electric field can extend into the adjacent tissues and cause ohmic heating of the tissues, although the object itself may not significantly heat (Nyenhuis (2018)).

Example: What lengths of wire are expected to be resonant at 1.5 T and 3 T? To compute what the resonant length of half the wavelength would be, it is necessary to establish the wavelength of the electromagnetic radiation in the body. In round numbers the frequency at 1.5 T is

$$\nu_0 = \gamma B_0 = (\frac{42.6\,\text{MHz}}{\text{T}})(1.5\text{T}) = 64\,\text{MHz} \quad (\text{A.15})$$

The wavelength in a given medium is related to the wavelength in vacuum by (Dempsey et al. (2001), ASTM F2182-19e2 (2019), Bennett et al. (2012))

$$\lambda_m = \frac{\lambda_0}{\sqrt{\varepsilon_r}} \quad (\text{A.16})$$

where λ_m is the wavelength in the medium such as tissue, λ_0 is the wavelength in vacuum, and ε_r is the dielectric constant of the medium. For air, which is close to that of vacuum, $\varepsilon_r = 1$ while that of tissue is approximately that of water at $\varepsilon_r = 80$.

Using $\lambda_0 = c/\nu_0$, the wavelength of rf radiation in air at 64 MHz calculates to

$$\lambda_m = \frac{\lambda_0}{\sqrt{\varepsilon_r}} = \frac{c}{\nu_0\sqrt{\varepsilon_r}} = \frac{(3 \times 10^8\, m\, s^{-1})}{(64 \times 10^6\, s^{-1})(\sqrt{1})}$$
$$= 4.69\, m = 469\, cm \tag{A.17}$$

The half wavelength value for a resonant length of wire in air would be $469\, cm/2 = 234.5\, cm$. Note, however, that in tissue

$$\lambda_m = \frac{\lambda_0}{\sqrt{\varepsilon_r}} = \frac{c}{\nu_0\sqrt{\varepsilon_r}} = \frac{(3 \times 10^8\, m\, s^{-1})}{(64 \times 10^6\, s^{-1})(\sqrt{80})}$$
$$= 0.52\, m = 52\, cm. \tag{A.18}$$

so that half wavelength for resonance for the same situation in the body is now $52\, cm/2 = 26\, cm$. The resonant length at 3 T would be half of that, or $13\, cm$. At 7 T it would be $5.6\, cm$ (Dempsey et al. (2001)). There is a considerable amount of uncertainty related to factors such as variable dielectric constant, and the resonant length at 1.5 T might vary from 23 to 30 cm and from 12 to 15 at 3 T. Note that the length does not have to be stretched out and could be in other configurations. The configuration and position in the body have been shown to affect the degree of heating (Mattei et al. (2008)). Also note that because of resonant length considerations it is possible for a lead to be dangerous at 1.5 T but not at 3 T (ACR Committee on MR Safety (2020)).

In cases where it is necessary to have external objects within the area exposed to the volume of the rf, strategies to mitigate risk include the placement of insulation between the object and patient, placing the lead as far from the inner walls of the bore as possible, and placement of cold compresses or ice packs to any areas of the object contacting the patient (ACR Committee on MR Safety (2020)).

Anthropogenic Gadolinium. Gadolinium found in the environment which may have accumulated from excreted gadolinium contrast agents (Rogowska et al. (2018)).

Anxiety. Feelings of nervousness, fear, apprehension, and worry. In the setting of MRI, this may be related to worry about what a test might show, worry about claustrophobia, or discomfort from the test. TJC has emphasized management of anxiety and claustrophobia as an important part of MRI, and the ACR safety manual recommends implementation of standard operating procedures and policies for management of claustrophobic and anxious patients (Spaeth and Gollub (2014), ACR Committee on MR Safety (2020)).

Architectural Issues. Facility design should support the four Zone concept as well as allow compliance with HIPAA and the Americans With Disabilities Act. As such the MR Zones must have patient access for those with disabilities, private patient interview and screening areas, a changing area for safe patient gowning and securing of valuables, a place in Zone II for physical screening with a ferromagnetic detector and/or handheld magnet with a location for quarantine of unsafe objects, physical securing and restriction of Zones III and IV, holding areas for patients in Zone III, and unrestricted lines of sight (Junk (2014), ACR Committee on MR Safety (2020)).

Artifact. Nonanatomic signal aberration in an MRI image. As this relates to MRI safety, artifacts are caused by implants, and an associated signal distortion caused by the difference in the magnetic susceptibility of the object and adjacent tissue (Faulkner (2016), ASTM F2119-07 (2013)).

Artifact Testing. In regard to MRI safety, a standardized test for determining the amount of image artifact an implant will generate under a given set of scan parameters (McRobbie (2020)).

(*MRSE*) Described in ASTM Standard F2119-07 (ASTM F2119-07 (2013)). Images are obtained using a prescribed set of spin echo and gradient echo sequences. The object is scanned with a reference object such as a nylon rod with the object and reference immersed in a CuSO4 solution. The artifact is measured in mm.

Asphyxiation. Deprivation of oxygen which may lead to unconsciousness and death. In the event of a quench breach, discharge of cryogen into the room may result in a decrease in oxygen levels which may be clinically significant. Symptoms and signs include tachypnea, confusion, headache, lightheadedness, nausea, seizure activity, and ultimately death (ACR Committee on MR Safety (2020)).

ASRT. American Society of Radiologic Technologists.

ASRT 2018 Radiologic Technologist Best Practices for MR Safety. A publication from the American Society of Radiologic Technologists on best practices for MRI safety (Clausen et al. (2018)). This document reaffirms a number of the ACR guidelines and includes: continuous supervision of Non-MR personnel in Zones III and IV, requirement that nonemergent patients be screened by two MR personnel at least one of which is Level 2 with verbal/interactive screening, recommendation for facility provided attire prior to entry to Zone III, safety screening documentation as part of the permanent medical record, hospital and community emergency responders as well as hospital staff should be provided yearly training offered by the facility, and finally service animals should not be allowed in Zone IV.

ASTM. See American Society for Testing and Materials International.

Asynchronous Pacing Mode. (*MRMD*) A pacing mode in which no sensing or inhibition occurs (Indik et al. (2017), Muthalaly et al. (2018)). In the three letter code system these are AOO, VOO, and DOO. During an MRI involving a pacemaker or ICD, the device is often programmed to an asynchronous mode. The rate should be programmed at a rate above the intrinsic heart rate to avoid the theoretical risk of an R on T excitation and associated induction of a potentially lethal tachyarrhythmia.

Atypical MR Environments. Systems which have MRI scanner components and operate outside the setting of conventional diagnostic imaging facilities. These facilities include intraoperative MRI, interventional MRI, positron emission tomography PET-MRI, and MRI-guided radiation therapy (Greenberg et al. (2019), ACR Committee on MR Safety (2020)). Individuals other than the traditional MR technologist or MR radiologist such as surgeons and other operating room personnel, radiation therapy personnel, nuclear medicine technologists, and others may work in an MR environment. Nonetheless, the same principles and procedures apply in these atypical settings as apply in the conventional diagnostic or research setting. As with conventional environments, there should be appropriate evaluation and screening of personnel and equipment entering the MR environment. In addition to MRI safety, procedures for infection control and potentially radioactive materials must also be in place appropriate to the facility.

B

B. The magnetic flux density. This is colloquially referred to as the "magnetic field".

(*MRSE*) The magnetic flux density in free space is

$$\mathbf{B} = \mu_0 \mathbf{H} \tag{B.1}$$

where **B** is the magnetic flux density, μ_0 is the permeability of free space and **H** is the magnetic field.

The magnetic flux density in the presence of matter induces a magnetization **M** of the matter and is given by

$$\mathbf{B} = \mu_0(\mathbf{H} + \mathbf{M}) \tag{B.2}$$

In general, **H** and **M** are related by a magnetic susceptibility tensor $\boldsymbol{\chi}$

$$\mathbf{M} = \boldsymbol{\chi}\mathbf{H} \tag{B.3}$$

However, for simple linear materials the susceptibility can be described by a scalar χ_m

$$\mathbf{M} = \chi_m \mathbf{H} \tag{B.4}$$

where χ_m is the magnetic susceptibility and therefore

$$\mathbf{B} = \mu_0(\mathbf{H} + \chi_m \mathbf{H}) = \mu_0(1 + \chi_m)\mathbf{H} \\ = \mu_0 \mu_r \mathbf{H} = \mu \mathbf{H} \tag{B.5}$$

where $\mu_r = 1+\chi_m$ is referred to as the relative magnetic permeability and μ is referred to as the magnetic permeability (Jiles (2015), Zangwill (2013), Coey (2010)).

$\mathbf{B}_0$. The magnetic flux density of the static main magnetic field. This is colloquially referred to as the magnetic field, although the magnetic field is technically the name usually given to $\mathbf{H}_0$.

(*MRSE*) In analogy for the general case for $\mathbf{B}$, the magnetic flux density of the static magnetic field in the absence of matter is (Jiles (2015), Zangwill (2013), Coey (2010))

$$\mathbf{B}_0 = \mu_0 \mathbf{H}_0 \tag{B.6}$$

with the introduction of matter this becomes

$$\mathbf{B}_0 = \mu_0(\mathbf{H}_0 + \mathbf{M}) \tag{B.7}$$

as in the general case for $\mathbf{B}$. See $\mathbf{B}$.

$\mathbf{B}_1$. The magnetic flux density field of the applied radiofrequency used to excite the spins in the patient. The SAR depends upon the square of B_1.

(*MRSE*) A linearly polarized $\mathbf{B}_1$ field at frequency ω applied in the x-y plane in the scanner (lab) frame can be modeled as (Jin (1999), Hoult (2000), Ansorge and Graves (2016), Ilott and Jerschow (2019))

$$\mathbf{B}_{1,lab} = B_{1x}\widehat{\mathbf{x}}cos\left(\omega t + \alpha\right) + B_{1y}\widehat{\mathbf{y}}cos\left(\omega t + \beta\right) \tag{B.8}$$

where $\mathbf{B}_{1,lab}$ is the applied field in the laboratory frame, B_{1x} and B_{1y} are amplitude factors, ω is the angular frequency, and α and β are the phases of the applied field. $\widehat{\mathbf{x}}$ and $\widehat{\mathbf{y}}$ are the unit vectors along x and y in the laboratory frame.

The vector amplitudes can be written in phasor form as

$$\mathcal{B}_{1x} = B_{1x}e^{i\alpha} \tag{B.9}$$

and

$$\mathcal{B}_{1y} = B_{1y}e^{i\beta} \tag{B.10}$$

$\mathbf{B}_{1,lab}$ can be considered as the sum of two components

$$\mathbf{B}_{1,lab} = \mathbf{B}_{1ccw,lab} + \mathbf{B}_{1cw,lab} \tag{B.11}$$

where $\mathbf{B}_{1ccw,lab}$ and $\mathbf{B}_{1cw,lab}$ are, respectively, the fields rotating counterclockwise and clockwise with respect to an observer in the laboratory frame. In a right-handed coordinate system with $\mathbf{B}_0$ oriented along $+z$ (Collins (2016), Ilott and Jerschow (2019))

$$\mathbf{B}_{1ccw,lab} = \frac{1}{2}\Big(B_{1x}\widehat{\mathbf{x}}cos\,(\omega t+\alpha) + B_{1x}\widehat{\mathbf{y}}sin\,(\omega t+\alpha) \\ - B_{1y}\widehat{\mathbf{x}}sin\,(\omega t+\beta) + B_{1y}\widehat{\mathbf{y}}cos\,(\omega t+\beta)\Big) \tag{B.12}$$

and

$$\mathbf{B}_{1cw,lab} = \frac{1}{2}\Big(B_{1x}\widehat{\mathbf{x}}cos\,(\omega t+\alpha) - B_{1x}\widehat{\mathbf{y}}sin\,(\omega t+\alpha) \\ + B_{1y}\widehat{\mathbf{x}}sin\,(\omega t+\beta) + B_{1y}\widehat{\mathbf{y}}cos\,(\omega t+\beta)\Big) \tag{B.13}$$

Using the identities

$$cos\,a = Re\left[e^{ia}\right] \tag{B.14}$$

and

$$sin\,a = Re\left[-ie^{ia}\right] \tag{B.15}$$

as well as

$$\frac{a}{i} = -ia \tag{B.16}$$

the counterclockwise field may be written

$$
\begin{aligned}
\mathbf{B}_{1ccw,lab} &= \frac{1}{2}\Big(Re\left[B_{1x}\widehat{\mathbf{x}}e^{i(\omega t+\alpha)} - iB_{1x}\widehat{\mathbf{y}}e^{i(\omega t+\alpha)}\right] \\
&- Re\left[-iB_{1y}\widehat{\mathbf{x}}e^{i(\omega t+\beta)} + B_{1y}\widehat{\mathbf{y}}e^{i(\omega t+\beta)}\right] \Big) \\
&= \frac{1}{2}\Big(Re\left[B_{1x}e^{i(\omega t+\alpha)}\left(\widehat{\mathbf{x}} - i\widehat{\mathbf{y}}\right)\right] \\
&+ Re\left[iB_{1y}e^{i(\omega t+\beta)}\left(\widehat{\mathbf{x}} - i\widehat{\mathbf{y}}\right)\right] \Big) \\
&= \frac{1}{2} Re\left[\left(B_{1x}e^{i(\omega t+\alpha)} + iB_{1y}e^{i(\omega t+\beta)}\right)\left(\widehat{\mathbf{x}} - i\widehat{\mathbf{y}}\right)\right]
\end{aligned}
\tag{B.17}
$$

The amplitude portion is often written in phasor notation

$$
\begin{aligned}
B_{1ccw,lab} &= \frac{1}{2}\left(B_{1x}e^{i\alpha} + iB_{1y}e^{i\beta}\right) = \widetilde{B}_{1x} + i\widetilde{B}_{1y} \\
&= \frac{1}{2}\left(\mathcal{B}_{1x} + i\mathcal{B}_{1y}\right)
\end{aligned}
\tag{B.18}
$$

Similarly, $\mathbf{B}_{1cw,lab}$ can be written as

$$
\begin{aligned}
\mathbf{B}_{1cw,lab} &= \frac{1}{2}\Big(Re\left[B_{1x}\widehat{\mathbf{x}}e^{-i(\omega l+\alpha)} - (-i)B_{1x}\widehat{\mathbf{y}}e^{-i(\omega t+\alpha)}\right] \\
&+ Re\left[-iB_{1y}\widehat{\mathbf{x}}e^{-i(\omega t+\beta)} + B_{1y}\widehat{\mathbf{y}}e^{-i(\omega t+\beta)}\right] \Big) \\
&= \frac{1}{2}\Big(Re\left[B_{1x}e^{-i(\omega t+\alpha)}\left(\widehat{\mathbf{x}} - i\widehat{\mathbf{y}}\right)\right] \\
&+ Re\left[iB_{1y}e^{-i(\omega t+\beta)}\left(\widehat{\mathbf{x}} - i\widehat{\mathbf{y}}\right)\right] \Big) \\
&= \frac{1}{2} Re\left[\left(B_{1x}e^{-i(\omega t+\alpha)} + iB_{1y}e^{-i(\omega t+\beta)}\right)\left(\widehat{\mathbf{x}} - i\widehat{\mathbf{y}}\right)\right]
\end{aligned}
\tag{B.19}
$$

In similar fashion the amplitude portion can also be written in phasor notation (Hoult (2000), Collins (2016))

$$B_{1cw,lab} = \frac{1}{2}\left(B_{1x}e^{-i\alpha} + iB_{1y}e^{-i\beta}\right) = \frac{1}{2}\left(B_{1x}e^{i\alpha} - iB_{1y}e^{i\beta}\right)^* \\ = \left(\widetilde{B}_{1x} - i\widetilde{B}_{1y}\right)^* = \frac{1}{2}\left(\mathcal{B}_{1x} - i\mathcal{B}_{1y}\right)^* \tag{B.20}$$

Taking the simple example of a field in the laboratory (scanner) frame linearly polarized along the x-axis in the lab frame with $\alpha = 0$ and $B_{1x} = B_1$ ($B_{1y} = 0$), then (Brown et al. (2014))

$$\mathbf{B}_{1cw,lab,Lin} = \frac{B_1}{2}\left(\widehat{\mathbf{x}}cos\omega t - \widehat{\mathbf{y}}sin\omega t\right) = \frac{B_1}{2}\widehat{\mathbf{x}}' \tag{B.21}$$

where $\mathbf{x}'$ is the unit vector along x in the rotating frame. The field is time independent in the rotating frame.

For a circularly polarized field there are two linearly polarized fields 90° out of phase in the laboratory frame, for example (Brown et al. (2014))

$$\mathbf{B}_{1cw,lab,Cir} = B_1\left(\widehat{\mathbf{x}}cos\omega t - \widehat{\mathbf{y}}sin\omega t\right) = B_1\widehat{\mathbf{x}}' \tag{B.22}$$

Note that for a linearly polarized field only half of the $\mathbf{B}_1$ field is available to nutate the spins. Linear polarization therefore requires twice the power input and twice the SAR of circular polarization to achieve the same flip angle.

$\mathbf{B}_{1+}$. Component of $\mathbf{B}_1$ rotating in the same direction as spin precession.

(*MRSE*) $\mathbf{B}_{1+}$ is given by, for example, in the case of clockwise precession

$$\begin{aligned}\mathbf{B}_{1+} &= \mathbf{B}_{1cw,lab} \\ &= \frac{1}{2} Re\left[\left(B_{1x}e^{-i(\omega t+\alpha)} + iB_{1y}e^{-i(\omega t+\beta)}\right)(\widehat{\mathbf{x}} - i\widehat{\mathbf{y}})\right]\end{aligned} \tag{B.23}$$

where B_{1x} and B_{1y} are amplitude factors, ω is the frequency, and α and β are the phases of the $\mathbf{B}_1$ field. $\widehat{\mathbf{x}}$ and $\widehat{\mathbf{y}}$ are the unit vectors along x and y in the laboratory frame. See $\mathbf{B}_1$.

The amplitude in Eq. B.23 is often written in phasor notation (Hoult (2000), Collins (2016)), which for clockwise precession is

$$B_{1+} = \left(\widetilde{B}_{1x} - i\widetilde{B}_{1y}\right)^* = \frac{1}{2}\left(\mathcal{B}_{1x} - i\mathcal{B}_{1y}\right)^* \tag{B.24}$$

In the case of counterclockwise precession (Collins (2016))

$$\begin{aligned}\mathbf{B}_{1+} &= \mathbf{B}_{1ccw,lab} \\ &= \frac{1}{2} Re\left[\left(B_{1x}e^{i(\omega t+\alpha)} + iB_{1y}e^{i(\omega t+\beta)}\right)(\widehat{\mathbf{x}} - i\widehat{\mathbf{y}})\right]\end{aligned} \tag{B.25}$$

which in phasor notation is (Collins (2016))

$$B_{1+} = \widetilde{B}_{1x} + i\widetilde{B}_{1y} = \frac{1}{2}\left(\mathcal{B}_{1x} + i\mathcal{B}_{1y}\right) \qquad \text{(B.26)}$$

See $\mathbf{B}_1$. For the case of a circularly polarized field with $B_{1x} = B_{1y} = B_1$ and $\alpha = \beta = 0$, Eq. B.23 becomes (Brown et al. (2014), Ilott and Jerschow (2019))

$$\mathbf{B}_{1+} = \mathbf{B}_{1cw,lab,Cir} = B_1\left(\widehat{\mathbf{x}}cos\omega t - \widehat{\mathbf{y}}sin\omega t\right) \qquad \text{(B.27)}$$

$\mathbf{B}_{1+rms}$. Root mean squared value of the $\mathbf{B}_1$ field in the same direction as precession ($\mathbf{B}_{1+}$) averaged over 10 second periods. This is typically in the range of 1-10 μT. This may replace SAR as a better means to estimate patient rf exposure as it does not depend upon assumption of a patient model or individual patient factors such as weight. For a given pulse sequence with a given set of parameters using a given set of scanner hardware, B_{1+rms} is always the same. The parameters can be saved for future use. Some implant labeling specifies B_{1+rms} as one of the scan limits.

(*MRSE*) B_{1+rms} is given by (International Organization for Standardization (2018), McRobbie (2020))

$$B_{1+rms} = \sqrt{\frac{\int_0^T \left|\mathbf{B}_{1+}(t)\right|^2 dt}{T}} \qquad \text{(B.28)}$$

where $\mathbf{B}_{1+}(t)$ is the $\mathbf{B}_1$ in the direction of precession, averaged over the time period T.

For a circularly polarized $\mathbf{B}_1$ field such as (See $\mathbf{B}_{1+}$)

$$\mathbf{B}_{1cw,lab,Cir} = \mathbf{B}_{1+} = B_1\left(\widehat{\mathbf{x}}cos\omega t - \widehat{\mathbf{y}}sin\omega t\right) \tag{B.29}$$

In the case of a simple model pulse Eq. B.28 works out to $B_{1+rms} = B_1$ for a frequency on the order of 64 MHz. See $\mathbf{B}_1$.

The following proportionality may be used (McRobbie (2020))

$$B_{1+rms} \propto \alpha_p \sqrt{\frac{N_E N_S}{\tau_p T_R}} \tag{B.30}$$

where α_p is the flip and angle, N_E is the number of echoes, N_S is the number of slices, τ_p is the pulse time, and T_R is the repetition time.

Example: The calculated B_{1+rms} for a sequence with a refocussing pulse of 180-degrees and T_R of 4000 ms with 16 echoes is 5.0 μT. If the parameters are changed so that the refocussing pulse is 180-degrees and T_R is 4500 ms with 8 echoes what will the new B_{1+rms} be?

$$\begin{aligned}\frac{B_{1+rms_2}}{B_{1+rms_1}} &= \alpha_{p2}\sqrt{\frac{N_{E_2}N_{S_2}}{\tau_{p_2}T_{R_2}}} \Big/ \alpha_{p1}\sqrt{\frac{N_{E_1}N_{S_1}}{\tau_{p_1}T_{R_1}}} \\ &= 150\sqrt{\frac{8}{4500}} \Big/ 180\sqrt{\frac{16}{4000}} = 0.56\end{aligned} \tag{B.31}$$

so that $B_{1+rms_2} = 0.56 B_{1+rms_2} = (0.56)(5.0\,\mu\text{T}) = 2.8\,\mu\text{T}$.

Bandwidth. A range of frequencies. This may be in terms of the range of frequencies transmitted or received (Brown et al. (2014)).

Biot-Savart Law. (*MRSE*) The Law of Biot and Savart is a general empirical law governing the magnetic field produced by current densities. A knowledge of this law is useful for understanding the three dimensional distribution of magnetic fields, which is in turn relevant to MRI safety. In its general form the law is (Brown et al. (2014))

$$\mathbf{B}(\mathbf{r}) = \frac{\mu_0}{4\pi} \int_V \frac{\mathbf{J}(\mathbf{r}') \times (\mathbf{r} - \mathbf{r}')}{(r - r')^3} dV' \tag{B.32}$$

where $\mathbf{B}(\mathbf{r})$ is the magnetic flux density at the observation point $\mathbf{r}$, μ_0 is the permeability of free space, and $\mathbf{J}(\mathbf{r}')$ is the current density at the source point $\mathbf{r}'$. Using the definition of current density (see Current Density), for a thin circular current loop with current I

$$\mathbf{B}(\mathbf{r}) = \frac{\mu_0}{4\pi} \oint_C \frac{I d\mathbf{l}' \times (\mathbf{r} - \mathbf{r}')}{(r - r')^3} \tag{B.33}$$

In differential form this is

$$d\mathbf{B}(\mathbf{r}) = \frac{\mu_0}{4\pi} \frac{I d\mathbf{l}' \times (\mathbf{r} - \mathbf{r}')}{(r - r')^3} \tag{B.34}$$

For a current loop with radius R_c contained in the xy-plane with center on the z-axis, then

$$d\mathbf{l}' = -R_c d\phi(sin\phi\widehat{\mathbf{x}} - cos\phi\widehat{\mathbf{y}}) \tag{B.35}$$

and

$$\mathbf{r} - \mathbf{r}' = -R_c(cos\phi\widehat{\mathbf{x}} + sin\phi\widehat{\mathbf{y}}) + (z - z')\widehat{\mathbf{z}} \tag{B.36}$$

Combining gives

$$\frac{Id\mathbf{l}' \times (\mathbf{r} - \mathbf{r}')}{(r - r')^3} = \frac{IR_c d\phi(z - z')(cos\phi\widehat{\mathbf{x}} - sin\phi\widehat{\mathbf{y}})}{2\left[R_c^2 + (z - z')^2\right]^{3/2}} + \frac{IR_c^2 d\phi\widehat{\mathbf{z}}}{2\left[R_c^2 + (z - z')^2\right]^{3/2}} \tag{B.37}$$

Therefore for a circular current loop at z' the magnetic field along the z-axis is

$$\mathbf{B}(z) = \frac{\mu_0}{4\pi}\int_0^{2\pi} \frac{IR_c d\phi(z - z')(cos\phi\widehat{\mathbf{x}} - sin\phi\widehat{\mathbf{y}}) + IR_c^2 d\phi\widehat{\mathbf{z}}}{2\left[R_c^2 + (z - z')^2\right]^{3/2}} = \left[\frac{\mu_0 I R_c^2}{2\left[R_c^2 + (z - z')^2\right]^{3/2}}\right]\widehat{\mathbf{z}} \tag{B.38}$$

for a single current loop.

To achieve a longer range of spatial field homogeneity two coils may be placed at $z' = +s'$ and $z' = -s'$ so that in this case

$$\mathbf{B}_{s'}(z) = \left[\frac{\mu_0 I R_c^2}{2\left[R_c^2 + (z - s')^2\right]^{3/2}} + \frac{\mu_0 I R_c^2}{2\left[R_c^2 + (z + s')^2\right]^{3/2}}\right]\widehat{\mathbf{z}} \tag{B.39}$$

For the particular case of a Helmholtz coil configuration in which $s' = R_c/2$ there is a field given by

$$\begin{aligned}\mathbf{B}_h(z) = 4\mu_0 I R_c^2 &\left[\frac{1}{(5R_c^2 - 4Rz + 4z^2)^{3/2}}\right]\widehat{\mathbf{z}} \\ + 4\mu_0 I R_c^2 &\left[\frac{1}{(5R_c^2 + 4R_c z + 4z^2)^{3/2}}\right]\widehat{\mathbf{z}}\end{aligned} \tag{B.40}$$

Similarly, in an MRI scanner multiple separate coils are placed in a configuration to provide a homogeneous field within the bore of the scanner (Brown et al. (2014), Ansorge and Graves (2016)). As a simple model consider the following graph in Figure B.1 of the static field for a Helmholtz pair showing how the field varies with distance.

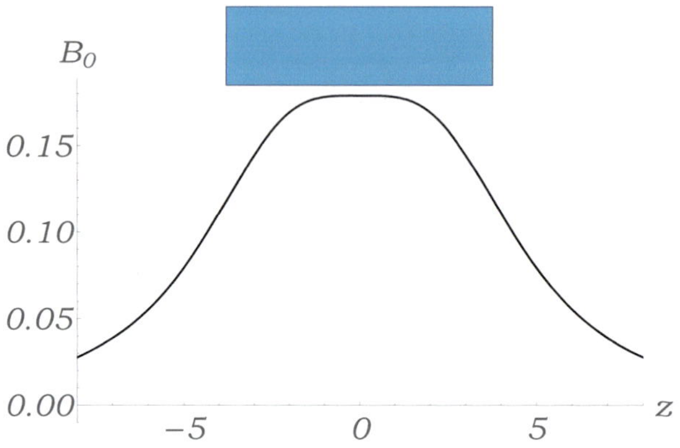

Figure B.1: Static field for a Helmholtz coil pair. The blue rectangle indicates the portion of the field as would be analogously present in the bore of an MRI scanner.

Birdcage Coil. A cylindrical rf coil which consists of rings on each end with multiple struts connecting the two rings to form a cylinder. Most body transmit coils are of this design. In addition, this configuration can be used as a transmit/receive coil (Collins (2016), Ansorge and Graves (2016)).

(*MRSE*) The birdcage coil can be modeled along the same lines taken by Jin (Jin (1999)) assuming an infinite hollow cylinder with radius R on which there is a surface current density (Becerra et al. (2006))

$$\mathbf{J}_s = J_0 sin(\omega t + \phi)\widehat{\mathbf{z}} \tag{B.41}$$

The magnetic field $\mathbf{B}$ on the cylinder can be found using the magnetic scalar potential Ψ (Jin (1999),

Ansorge and Graves (2016)) assuming a small birdcage coil in the quasistatic limit

$$\mathbf{B} = -\nabla\Psi \tag{B.42}$$

Taking the divergence of both sides gives

$$\nabla \cdot \mathbf{B} = \nabla \cdot \big(-\nabla\Psi \big) = -\nabla^2\Psi \tag{B.43}$$

Since $\nabla{\cdot}\mathbf{B} = 0$, this gives the Laplace equation

$$\nabla^2\Psi = 0 \tag{B.44}$$

The general solution of the Laplace equation in cylindrical coordinates if the potential does not vary along the z-axis is (Jin (1999), Knoepfel (2000))

$$\Psi(r, \phi, 0) = \sum_{m=-\infty}^{\infty} r^m \big(A_m cos\, m\phi + B_m sin\, m\phi \big) \tag{B.45}$$

Taking the gradient in cylindrical coordinates to obtain the magnetic field gives the general solution

$$\begin{aligned} \mathbf{B}(r, \phi, 0) &= -\nabla\Psi(r, \phi, 0) \\ &= -\widehat{\boldsymbol{r}} \sum_{m=-\infty}^{\infty} m r^{m-1} \big(A_m cos\, m\phi + B_m sin\, m\phi \big) \\ &+ \widehat{\boldsymbol{\phi}} \sum_{m=-\infty}^{\infty} m r^{m-1} \big(A_m sin\, m\phi - B_m cos\, m\phi \big) \end{aligned} \tag{B.46}$$

Now starting to apply the electromagnetic boundary conditions at the surface of the coil provides a field inside the coil and outside the coil, and noting that the term for $m = 0$ is zero. The field must go toward zero moving away from the coil and must be finite at the center of the coil at $r = 0$ so for $r < R$ (Jin (1999), Ansorge and Graves (2016))

$$\begin{aligned}\mathbf{B}_{r<R} =& \\ &-\widehat{\boldsymbol{r}}\sum_{m=1}^{\infty} mr^{m-1}\big(A_m cos\, m\phi + B_m sin\, m\phi\big) \\ &+\widehat{\boldsymbol{\phi}}\sum_{m=1}^{\infty} mr^{m-1}\big(A_m sin\, m\phi - B_m cos\, m\phi\big)\end{aligned} \tag{B.47}$$

and

$$\begin{aligned}\mathbf{B}_{r>R} =& \\ &-\widehat{\boldsymbol{r}}\sum_{i=-\infty}^{-1} mr^{m-1}\big(A_m cos\, m\phi + B_m sin\, m\phi\big) \\ &+\widehat{\boldsymbol{\phi}}\sum_{i=-\infty}^{-1} mr^{m-1}\big(A_m sin\, m\phi - B_m cos\, m\phi\big) \\ &=\widehat{\boldsymbol{r}}\sum_{i=1}^{\infty} mr^{-m-1}\big(A_{-m} cos\, m\phi - B_{-m} sin\, m\phi\big) \\ &-\widehat{\boldsymbol{\phi}}\sum_{i=1}^{\infty} mr^{-m-1}\big(-A_{-m} sin\, m\phi + B_{-m} cos\, m\phi\big)\end{aligned} \tag{B.48}$$

Use the electromagnetic boundary condition at the coil surface that

$$\widehat{\mathbf{n}} \times \left(\mathbf{B}_{r>R} - \mathbf{B}_{r<R}\right) = \mu_0 \mathbf{J}_s \tag{B.49}$$

where the unit vector $\widehat{\mathbf{n}}$ is normal to the surface of the cylindrical coil to obtain

$$\begin{aligned} \mu_0 \mathbf{J}_s &= \mu_0 J_0 sin(\omega t + \phi)\widehat{\mathbf{z}} \\ &= +\widehat{\boldsymbol{\phi}} \sum_{i=1}^{\infty} m r^{m-1}\left(A_m sin\, m\phi + B_m cos\, m\phi\right) \\ &- \widehat{\boldsymbol{\phi}} \sum_{m=1}^{\infty} m r^{m-1}\left(A_m sin\, m\phi - B_m cos\, m\phi\right) \end{aligned} \tag{B.50}$$

To find the coefficients, it can be recognized that this can be written as a standard Fourier series to give (Riley et al. (2006), Griffiths (2017))

$$f(r,\phi) = \sum_{m=1}^{\infty} a_m cos m\phi + b_m sin m\phi \tag{B.51}$$

where

$$a_m = m r^{m-1} B_m + m r^{-m-1} B_{-m} \tag{B.52}$$

and

$$b_m = -m r^{m-1} A_m + m r^{-m-1} A_{-m} \tag{B.53}$$

Using the standard formulas for finding the Fourier coefficients (Riley et al. (2006), Griffiths (2017))

$$a_n = \frac{1}{\pi}\int_{-\pi}^{\pi} f(r,\phi)cosn\phi d\phi \tag{B.54}$$

and

$$b_n = \frac{1}{\pi}\int_{-\pi}^{\pi} f(r,\phi)sinn\phi d\phi \tag{B.55}$$

with the orthogonality properties of sine and cosine, only the $n = m = 1$ terms are non-zero and together with $r = R$ at the boundary, the Fourier series is

$$\begin{aligned} f(r,\phi) &= \mu_0 J_0 sin(\omega t + \phi) = \mu_0 J_0\big(sin\omega t cos\phi + cos\omega t sin\phi\big) \\ &= a_1 sin\phi + b_1 sin\phi \\ &= \big(\frac{B_{-1}}{r^2} + B_1\big)cos\phi + \big(\frac{A_{-1}}{r^2} - A_1\big)sin\phi \end{aligned} \tag{B.56}$$

implying that

$$\frac{B_{-1}}{R^2} + B_1 = \mu_0 J_0 sin\omega t \tag{B.57}$$

and

$$\frac{A_{-1}}{R^2} - A_1 = \mu_0 J_0 cos\omega t \tag{B.58}$$

Now using the other boundary condition

$$\widehat{\mathbf{n}} \cdot \big(\mathbf{B}_{r>R} - \mathbf{B}_{r<R}\big) = 0 \tag{B.59}$$

so that

$$
\begin{aligned}
&-\hat{\boldsymbol{r}}\sum_{m=1}^{\infty} mr^{m-1}\big(A_m cos\, m\phi + B_m sin\, m\phi\big) \\
&= -\hat{\boldsymbol{r}}\sum_{i=-\infty}^{-1} mr^{m-1}\big(A_m cos\, m\phi + B_m sin\, m\phi\big) \\
&= -\hat{\boldsymbol{r}}\sum_{i=1}^{\infty} -mr^{-m-1}\big(A_{-m} cos\,\phi - B_{-m} sin\,\phi\big)
\end{aligned} \tag{B.60}
$$

together with the condition that only $m = 1$ terms are non-zero

$$
-A_1 cos\,\phi + B_1 sin\,\phi = \frac{A_{-1}}{r^2} cos\,\phi - \frac{B_{-1}}{r^2} sin\,\phi \tag{B.61}
$$

implies

$$
A_1 = -\frac{A_{-1}}{R^2} \tag{B.62}
$$

and

$$
B_1 = \frac{B_{-1}}{R^2} \tag{B.63}
$$

From Eqs B.57, B.58, B.62, and B.63

$$
B_1 = \frac{\mu_0 J_0}{2} sin\omega t \tag{B.64}
$$

and

$$B_{-1} = \frac{R^2 \mu_0 J_0}{2} sin\omega t \tag{B.65}$$

and

$$A_1 = -\frac{\mu_0 J_0}{2} cos\omega t \tag{B.66}$$

and

$$A_{-1} = \frac{R^2 \mu_0 J_0}{2} sin\omega t \tag{B.67}$$

Therefore, using Eq. B.47 inside the coil $r < R$

$$\begin{aligned} &\mathbf{B}_{r<R} = \\ &\frac{\mu_0 J_0}{2}\widehat{\boldsymbol{r}}\big(cos\omega t cos\phi - sin\omega t sin\phi\big) \\ &- \frac{\mu_0 J_0}{2}\widehat{\boldsymbol{\phi}}\big(cos\omega t sin\phi + sin\omega t cos\phi\big) \end{aligned} \tag{B.68}$$

and upon transforming the field to Cartesian coordinates gives a circularly polarized field (See $\mathbf{B}_1$)

$$\begin{aligned} \mathbf{B}_{r<R} &= \mathbf{B}_1 = \mathbf{B}_{1cw,lab,Cir} \\ &= \frac{\mu_0 J_0}{2}\big(\widehat{\mathbf{x}} cos\omega t - \widehat{\mathbf{y}} sin\omega t\big) \\ &= B_1\big(\widehat{\mathbf{x}} cos\omega t - \widehat{\mathbf{y}} sin\omega t\big) \end{aligned} \tag{B.69}$$

Using a similar analysis and

$$\mathbf{J}_s = J_0 cos\omega t sin(\phi)\widehat{\mathbf{z}} \tag{B.70}$$

then the linearly polarized field

$$\begin{aligned}\mathbf{B}_{r<R} &= \mathbf{B}_1 = \mathbf{B}_{1,lab} \\ &= \frac{\mu_0 J_0}{2}\widehat{\mathbf{x}}cos\omega t \\ &= B_1\widehat{\mathbf{x}}cos\omega t\end{aligned} \tag{B.71}$$

is obtained. If it is assumed $\omega = 0$ then

$$\mathbf{B}_{r<R} = B_1\widehat{\mathbf{x}} \tag{B.72}$$

is obtained as in Jin (Jin (1999)) and Ansorge and Graves (Ansorge and Graves (2016)).

Body Habitus. Body physical form. This impacts MRI safety to the extent that patients with a larger body habitus will tend to heat more from MRI. In addition, patients with less surface area to volume ratio will be less efficient at cooling (McRobbie et al. (2017)). Because a larger body habitus will potentially be more likely to have a body part close to the inner surface of the bore of the magnet, near field burns may be more likely in this population. See Specific Absorption Rate, Near Field Burn.

Height may also affect safety as a taller patient has a greater chance that a foreign object or implant may be sufficiently far from potentially dangerous regions of the scanner to allow safe imaging. See Patient Height.

Body Transmit Coil. An rf coil built into the wall of the scanner which excites much of volume in the central bore. This is usually a birdcage coil design. This is used in conjunction with a local receive only coil to detect the MRI signal from the patient (Collins (2016), Brown et al. (2014)). While the built in transmit coil excites the entire volume over its length, the field quickly falls off outside the coil. See Field Plot.

Breastfeeding and Contrast. The 2018 ACR Manual on Contrast Media (ACR Committee on Drugs and Contrast Media (2020)) indicates that less than 0.04% of the gadolinium contrast dose administered to the mother is excreted in her breastmilk and that less than 1% of that is absorbed by the infant's GI tract so that less than 0.0004% of the original dose is absorbed by the infant.

Given the small amount of gadolinium absorbed there is no recommendation to stop breastfeeding for gadolinium administration (ACR Committee on Drugs and Contrast Media (2020)). At the mother's discretion, she may elect to pump and dump breastmilk for 12 to 24 hours. Prior to an elective examination the mother may want to pump additional milk to use for the 12 to 24 hours afterward. There is no indication that stopping longer than 24 hours is of value. Gadolinium in breast milk could potentially alter the taste of the milk, which could affect feeding (ACR Committee on Drugs and Contrast Media (2020)).

Broken Leads. If a lead on a conditional device is broken, the device no longer meets the safety conditions to scan the implant, and patient injury could be more likely (McRobbie (2020)).

Burn. In the setting of MRI safety, most often refers to a high temperature thermal injury, and one of the most common adverse events (Delfino et al. (2019)). Burns in MRI are thought to occur by multiple mechanisms including hot spots, near field proximity burns, inductive loop burns, antenna effect burns, reflective burns, and resonant LC circuit burns. Note that the patient may not feel the burn as it is happening in MRI, but may develop symptoms and skin changes hours later. Hence the patient may not be able to warn the technologist prior to the injury (Bennett et al. (2012), Dempsey et al. (2001), Dempsey and Condon (2001), Schaefer (2014), Tsai et al. (2015)). Low temperature thermal injury (frostbite) can occur from exposure to cryogens. See Frostbite.

C

Capacitance (C). (*MRSE*) Ability of a system to store an electric charge. Measured in Farads (Halliday et al. (2014)).

Capacitor. (*MRSE*) An electrical device which stores electric charge (Halliday et al. (2014)).

Cardiac Implantable Electronic Device (CIED). Active implant designed to monitor and/or support the electrical function of the heart. Examples: Pacemakers, ICDs, and loop recorders (Indik et al. (2017)).

Cartesian Coordinates. Traditional coordinate system utilizing x, y, and z coordinates (Griffiths (2017)).

cgs Units. Older system of units based on three base units of centimeters, seconds, and grams. A derived unit commonly used relevant to MRI safety is the Gauss (G).

Chelate. A compound with a usually organic ligand binding a metal ion at two or more sites. To form a chelate. Gadolinium contrast agents are chelates. Chelates relevant to MRI can be classified as linear or macrocyclic in terms of their three dimensional structure and ionic or nonionic with respect to their binding mechanism (ACR Committee on Drugs and Contrast Media (2020), Rogowska et al. (2018)).

Chronaxie (C_h). The time required for an electric current to double the rheobase in nerve stimulation (McRobbie (2014)). See Strength Duration Curve.

Chronic Kidney Disease (CKD). Chronic renal dysfunction, which is a risk factor for developing nephrogenic systemic sclerosis (NSF) from gadolinium based contrast media (GBCM) depending on degree of dysfunction.

(*MRMD*) Chronic kidney disease (CKD) is defined by (Simon et al. (2011), ACR Committee on Drugs and Contrast Media (2020)):

Either

Kidney damage lasting at least 3 months regardless of GFR based on abnormal biopsy, blood, urine, or radiological tests

Or

GFR $< 60\, m\text{L}/min/1.73\, m^2$ for 3 or more months regardless of other testing.

CKD is divided into stages CKD1-5 defined as follows:

CKD1: Kidney damage and GFR $\geq 90\, m\text{L}/min/1.73\, m^2$.
CKD2: Kidney damage and GFR 60-89 $m\text{L}/min/1.73\, m^2$.
CKD3: GFR 30-59 $m\text{L}/min/1.73\, m^2$.
CKD4: GFR 15-29 $m\text{L}/min/1.73\, m^2$.
CKD5: GFR $< 15\, m\text{L}/min/1.73\, m^2$.

The ACR suggests a lower limit GFR of $30\, m\text{L}/min/1.73\, m^2$ for gadolinium contrast administration without special precautions in patients with chronic kidney disease (ACR Committee on Drugs and Contrast Media (2020)).

CIED. See Cardiac Implantable Electronic Device.

CKD. See Chronic Kidney Disease.

Claustrophobia. Distress when in a close or enclosed space such as an MRI scanner (Spaeth and Gollub (2014)). The 2020 ACR safety manual recommends implementation of standard operating procedures and policies for management of claustrophobic and anxious patients (ACR Committee on MR Safety (2020)).

Clothing. Some clothing contains electrically conductive material, such as silver or copper for antimicrobial activity. This material can potentially cause burns (Delfino et al. (2019)). Clothing product labeling will not reliably exclude the presence of such material, and therefore gowning to skin with MR Safe gowns or scrubs supplied by the imaging site is recommended (ACR Committee on MR Safety (2020)).

Coil Polarization. The orientation of the magnetic $\mathbf{B}_1$ field, for example linearly or circularly polarized. Circularly polarized (also called quadrature) coils deliver less power and have better signal to noise than linear coils (Brown et al. (2014)).

(*MRSE*) For a linearly polarized coil in the laboratory frame with frequency ω (Brown et al. (2014))

$$\mathbf{B}_{1lab,Lin} = \widehat{\mathbf{x}} B_1 cos\omega t \tag{C.1}$$

where $\widehat{\mathbf{x}}$ is the unit vector along x in the laboratory frame. See $\mathbf{B}_1$.

To an observer in the laboratory frame, the field in the clockwise rotating frame is

$$\mathbf{B}_{1cw,lab,Lin} = \frac{B_1}{2}\left(\widehat{\mathbf{x}}cos\omega t - \widehat{\mathbf{y}}sin\omega t\right) = \frac{B_1}{2}\widehat{\mathbf{x}'} \tag{C.2}$$

where $\widehat{\mathbf{y}}$ is the unit vector along y in the laboratory frame. $\mathbf{x}'$ is the unit vector along x in the rotating frame. To a spin in the rotating frame the field appears static along x with an amplitude of $B_1/2$.

For a circularly polarized field there are two linearly polarized fields added together in the laboratory frame to give in the clockwise rotating frame (Brown et al. (2014))

$$\mathbf{B}_{1cw,lab,Cir} = B_1\left(\widehat{\mathbf{x}}cos\omega t - \widehat{\mathbf{y}}sin\omega t\right) = B_1\widehat{\mathbf{x}'} \tag{C.3}$$

Note that for a linearly polarized field only half of the $\mathbf{B}_1$ field is available to nutate the spins. Linear polarization therefore requires twice the power input and twice the SAR of circular polarization to achieve the same flip angle. In addition, the linearly polarized coil only provides $\sqrt{2}$ the signal to noise of a circularly polarized coil (Brown et al. (2014)). See $\mathbf{B}_1$.

Competitive Pacing. Cardiac pacing can compete with the intrinsic rhythm. This is considered a contraindication to MRI for some conditional devices (Indik et al. (2017)).

Conduction. A mechanism of cooling in MRI. Movement of heat from atomic vibrations due to two bodies in direct contact (Halliday et al. (2014)).

Example: Ice pack used as a heat sink over a tattoo.

Conductivity (σ_c). Property of a substance indicating how well it conducts electricity. Inverse of the resistivity, $\sigma_c = 1/\rho_c$ (Halliday et al. (2014)). Measured in Siemens/meter (S/m).

Conductor. A substance through which charge moves freely (Halliday et al. (2014)). Example : Copper.

Contrast Reaction. An adverse physiological response following the administration of contrast media. These can be subdivided into allergic and non-allergic reactions. In addition, there is a reported effect of Gadoxetate disodium (Eovist) in which patients are unable to suspend respiration after injection (ACR Committee on Drugs and Contrast Media (2020)).

Convection. A mechanism of cooling in MRI. Movement of heat due to flow of fluid or gas (Halliday et al. (2014)). Example: Fan in MRI.

Cooling. Mechanisms of patient cooling include convection, conduction, radiation, and evaporation (Halliday et al. (2014)).

Coulomb (C). (*MRSE*) A unit of electrical charge. $1\,C = 1$ A-s (Halliday et al. (2014)).

Cryogen. A substance used for producing very low temperature cooling. In a superconducting MRI scanner, liquid helium and sometimes additional liquid nitrogen are used to cool the main coil. These boil off over time and have to be periodically refilled (Collins (2016), Faulkner (2016)). Newer designs use only liquid helium and have very little boil off.

MRI safety issues may arise during a refill of the cryogens or during a quench if cryogens escape during a refill or during a quench breach. For this reason refills should only be performed by qualified individuals with only essential personnel in the area during the refill (ACR Committee on MR Safety (2020)). See Cryogen Hazards.

Cryogen Hazards. Cryogen hazards include ruptured eardrums from increased pressure in the event of a quench breach, asphyxiation in the event of a quench breach from oxygen levels being decreased by escaping cryogen gas, room entrapment from increased pressure in the event of a quench breach, fire related to concentration of liquified oxygen, explosion if there is a quench pipe blockage during a quench, and frostbite from direct exposure of skin to cold (ACR Committee on MR Safety (2020)).

Curie Point. See Curie temperature.

Curie Temperature. Temperature above which a substance loses its magnetism (Coey (2010)).

Current (I). Charge motion with respect to time (Halliday et al. (2014)). SI unit is the Ampere (A).

Current Density (J). (*MRSE*) Vector quantity of current per area with units of A/m^2 (Zangwill (2013)).

The relationship of current I to current density J is

$$I = \int_S \mathbf{J} \cdot d\mathbf{A} \tag{C.4}$$

Also see Ohm's Law.

Cylindrical Coordinates. (*MRSE*) Coordinate system based on r, θ, and ϕ base vectors. The relationship of Cartesian to cylindrical coordinates is given by (Griffiths (2017))

$$\widehat{\boldsymbol{r}} = cos\phi\widehat{\mathbf{x}} + sin\phi\widehat{\mathbf{y}} \tag{C.5}$$

$$\widehat{\boldsymbol{\phi}} = -sin\phi\widehat{\mathbf{x}} + cos\phi\widehat{\mathbf{y}} \tag{C.6}$$

$$\widehat{\mathbf{z}} = \widehat{\mathbf{z}} \tag{C.7}$$

and

$$\widehat{\mathbf{x}} = cos\phi\widehat{\boldsymbol{r}} - sin\phi\widehat{\boldsymbol{\phi}} \tag{C.8}$$

$$\widehat{\mathbf{y}} = sin\phi\widehat{\boldsymbol{r}} + cos\phi\widehat{\boldsymbol{\phi}} \tag{C.9}$$

$$\widehat{\mathbf{z}} = \widehat{\mathbf{z}} \tag{C.10}$$

In Cartesian coordinates the unit vectors are given by $\widehat{\mathbf{x}}$, $\widehat{\mathbf{y}}$, and $\widehat{\mathbf{z}}$. In cylindrical coordinates, the unit vectors are denoted $\widehat{\boldsymbol{r}}$, $\widehat{\boldsymbol{\phi}}$, and $\widehat{\mathbf{z}}$.

D

dB. See Decibel.

dB(A). See A-Weighting.

$d\mathbf{B}/dt$. The change in **B** with time. In terms of MR safety, a change in the magnetic field with respect to time induces currents which can lead to heating and physiological effects through Faraday's Law as it applies to time varying rf and gradient fields or movement through the main static field. For the time varying gradients, the maximum values of dB/dt occur near the ends of the gradient coils near the ends of the bore, with the value of dB/dt is zero at the isocenter. For dB/dt due to rf fields, this is confined to the volume of the coil, with a maximum near the isocenter. In addition, dB/dt may also arise by movement of the body through the spatially varying static field.

Effects from dB/dt relevant to MR safety include symptoms of vertigo and nausea from movement in the field, peripheral nerve stimulation giving rise to magnetophosphenes and paresthesias from the switched applied gradient field, acoustic noise arising from the switching gradients, and magnetic induction of currents through Faraday's Law which leads to currents and heating. Although there is a theoretical risk of induced arrhythmias, the maximum dB/dt produced by current clinical systems is believed to be well below the threshold for arrhythmia induction (McRobbie (2014)). Induced currents within implants may cause heating or malfunction and/or damage to the implant.

($MRSE$) In terms of the magnitude of the time varying applied gradients, for example $G_{z_x} = dB_z/dx$, (Brown et al. (2014))

$$B_z = G_{z_x} x \tag{D.1}$$

and

$$\frac{dB_z}{dt} = \frac{dG_{z_x}}{dt} x \tag{D.2}$$

and for the maximum value

$$\frac{dB_{z\,Max}}{dt} = \frac{G_{z_x\,Max}}{t_r} x \tag{D.3}$$

so that the maximum dB/dt depends on the maximum gradient strength $G_{z_x\,Max}$ divided by the minimum rise time, t_r, as well as the position x. In order to have higher performance gradients with shorter rise times and higher gradient maximum values, there will be an increase in dB/dt.

In terms of rf excitation, a time varying linearly polarized rf field in the laboratory frame $\mathbf{B}_{1lab,Lin}$ has a field

$$\mathbf{B}_{1lab,Lin} = B_1 cos\omega_0 t\widehat{\mathbf{x}} \tag{D.4}$$

so that

$$\frac{d\mathbf{B}_{1lab,Lin}}{dt} = -B_1\omega_0 sin\omega_0 t\widehat{\mathbf{x}} \tag{D.5}$$

Note that the peak amplitude of dB/dt depends upon the amplitude of the applied rf as well as the the frequency ω_0 and hence B_0.

For motion in the static field with an associated spatial gradient the following can be considered

$$\frac{dB_0}{dt} = \frac{dB_0}{dz}v = \frac{dB_0}{dz}\frac{dz}{dt} \tag{D.6}$$

where $v = dz/dt$ is the velocity of motion along z through the static field gradient dB_0/dz.

Example: A technologist moves through a spatial gradient of 4 T/m with a speed of 1.5 m/s. What dB/dt is experienced?

$$\frac{dB}{dt} = \left(\frac{4\mathrm{T}}{s}\right)\left(\frac{1.5m}{s}\right) = 6\,\mathrm{T}/s \tag{D.7}$$

which is sufficiently great to cause symptoms.

$d\mathbf{B}/dz$. Spatial gradient. The change in **B** with distance. The magnetic displacement force depends upon the spatial gradient. See Spatial Gradient.

Decibel (dB). A measure widely used in acoustics as a measure of loudness (Halliday et al. (2014)).

(*MRSE*) Base 10 logarithm of the ratio between two numbers multiplied by a factor depending on the type of measurement. In MRI safety, the parameter of interest is the sound pressure level from gradient noise (McJury (2014)). The decibel level is calculated from

$$\mathrm{dB} = 20\log_{10}\left(\frac{\mathrm{P}}{\mathrm{P}_0}\right) \tag{D.8}$$

where $\mathrm{P}_0 = 2 \times 10^{-5}\,\mathrm{Pa}$, which is the lowest pressure

detectable by the human ear. Pa denotes pressure in pascals.

Deep Sedation. Sedation level in which patients are not easily aroused but respond purposefully to painful or repeated stimulation. Ventilatory function may be impaired (Reinking (2014)).

Defibrillator. Electronic device which applies electric energy to the heart to reestablish heart rhythm. These may be external or internal such as an ICD (Indik et al. (2017)).

Demagnetizing Factor (N). (*MRSE*) When an object is placed in a magnetic field there is a magnetic field induced in the object which typically opposes the applied field (Zangwill (2013)). This is important in MR safety regarding the degree to which an object can become magnetized and experience forces in the MR environment. The demagnetizing field $\mathbf{H}_d$ for an ellipsoid takes the form

$$\mathbf{H}_d = -\mathbf{N} \cdot \mathbf{M} \tag{D.9}$$

where $\mathbf{M}$ is the magnetization of the object and $\mathbf{N}$ is the demagnetizing tensor. In its principal axis system the individual diagonal tensor elements N_{ii} are known as demagnetizing factors (Cullity and Graham (2011), Zangwill (2013), Zélis et al. (2017)). See Demagnetizing Tensor.

Demagnetizing Field ($\mathbf{H}_d$). (*MRSE*) When an object is placed in a magnetic field $\mathbf{H}_0$ there is a magnetic field induced in the object which typically opposes the applied field. This is important in MR safety regarding the degree to which an object can become magnetized and experience forces in the MR environment. The demagnetizing field $\mathbf{H}_d$ for an ellipsoid takes the form (Zangwill (2013))

$$\mathbf{H}_d = -\mathbf{N} \cdot \mathbf{M} \tag{D.10}$$

where $\mathbf{M}$ is the magnetization of the object and $\mathbf{N}$ is the demagnetizing tensor.

The total field $\mathbf{H}_{in}$ inside a body in a magnetic field is

$$\mathbf{H}_{in} = \mathbf{H}_0 + \mathbf{H}_d = \mathbf{H}_0 - \mathbf{N} \cdot \mathbf{M} \tag{D.11}$$

Demagnetizing Tensor ($\mathbf{N}$). (*MRSE*) A demagnetizing tensor in general must be used to describe the effect of an external field on an object and its associated demagnetizing field (Zangwill (2013)). See Demagnetizing Field. The demagnetizing tensor is diagonal in its principle axis system and is

$$\mathbf{N} = \begin{pmatrix} N_{xx} & 0 & 0 \\ 0 & N_{yy} & 0 \\ 0 & 0 & N_{zz} \end{pmatrix} \tag{D.12}$$

For an ellipsoid the elements of the demagnetizing tensor N_{ii} in its principle axis system depend on the dimensions of the ellipsoid, l_a, l_b, and l_c with two of the dimensions equal, and are given by N_a and N_c as follows (Cullity and Graham (2011)):

For $l_a = l_b < l_c$ and $\zeta = l_c/l_a$ (prolate spheroid)

$$N_c = \frac{1}{\zeta^2 - 1}\left[\frac{\zeta}{\sqrt{\zeta^2 - 1}} ln\left(\zeta + \sqrt{\zeta^2 - 1}\right) - 1\right] \tag{D.13}$$

and

$$N_a = N_b = \frac{1 - Nc}{2} \tag{D.14}$$

For a long thin rod

$$N_c \approx \frac{1}{\zeta^2}\left[ln(2\zeta) - 1\right] \tag{D.15}$$

and

$$N_a = N_b \approx \frac{1}{2} \tag{D.16}$$

For $l_a < l_b = l_c$ and $\zeta = l_c/l_a$ (oblate spheroid)

$$N_c = N_b = \frac{1}{2(\zeta^2 - 1)}\left[\frac{\zeta^2}{\sqrt{1 - \zeta^2}} arcsin\left(\frac{\sqrt{\zeta^2 - 1}}{\zeta}\right) - 1\right] \tag{D.17}$$

and

$$N_a = 1 - 2N_c \tag{D.18}$$

For a long thin disk

$$N_c = N_b \approx \frac{\pi}{4\zeta} - \frac{1}{2\zeta^2} \approx \frac{\pi}{4\zeta} \tag{D.19}$$

and

$$N_a \approx 1 - \frac{\pi}{2\zeta} - \frac{1}{\zeta^2} \approx 1 - \frac{\pi}{2\zeta} \tag{D.20}$$

For $\zeta = l_c/l_a = 1$ (sphere)

$$N_a = N_b = N_c = \frac{1}{3} \tag{D.21}$$

Note that in all cases

$$N_a + N_b + N_c = 1 \tag{D.22}$$

An example used in multiple calculations in this book is for a prolate spheroid with the long axis along the x-axis so that $N_{xx} = N_c$ and $N_{yy} = N_{zz} = N_a = N_b$.

Density (ρ). A measure of a property per unit volume. For example the mass density is $\rho = \text{g}/cc$ (Halliday et al. (2014)). Density may also refer to a property per unit area such as surface charge density $\sigma_q = C/m^2$ (Zangwill (2013)).

Diamagnetic. A diamagnetic substance has randomly oriented atomic dipoles which become partially aligned against an external magnetic field and therefore exhibits a weakly repulsive force. The alignment disappears when the external field is removed. These substances have a magnetic susceptibility less than 1, typically on the order of -10^{-5}, and do not experience significant displacement force or torque from the static field. Example: Silver (Jiles (2015)). See Magnetic Susceptibility.

Diaphragmatic Pacing. A transvenous cardiac pacer may cause diaphragmatic stimulation as well as cardiac stimulation, which typically represents a contraindication to scanning.

Dielectric. A substance through which charge does not freely move (Griffiths (2017)). An insulator.

Dielectric Constant (ε_r). The relative electric permittivity. The wavelength of an electromagnetic wave in a medium such as tissue depends on the dielectric constant of the medium. This is important in the degree of heating which may occur during MRI. $\varepsilon_r = 1 + \chi_E$, where χ_E is the electric susceptibility (Zangwill (2013)).

Dipole. A pair of electric or magnetic poles (electric or magnetic charges) separated by some distance (Halliday et al. (2014), Zangwill (2013)).

Dipole Field. Electric or magnetic field generated around an electric or magnetic dipole (Zangwill (2013)).

Dipole Force. Force exerted by an electric or magnetic dipole (Zangwill (2013)).

Displacement Force. See Magnetic Displacement Force.

Displacement Force Testing. Assessment of the magnetic attractive force on an object. In basic terms the object is suspended from a string at the entrance to the bore of the MRI scanner. If there is a deflection less than 45 degrees from vertical, the magnetic force is less than the force of gravity, and the object is not expected to pose a risk from the displacement force.

(*MRSE*) Described in ASTM Standard F2052 -15 (ASTM F2052-15 (2015)). Test must be carried out on a horizontal bore system with horizontal field parallel to the bore at a location within 20% of the maximum value of the spatial gradient. Suspend object from string along protractor at the entrance of the bore on axis of the bore. Record the location, the magnetic flux density, and the spatial gradient. Measure the deflection angle α between the device and the vertical to the nearest degree. Calculate the magnitude of the magnetic force F_m using

$$F_m = F_g \tan\alpha = mg \tan\alpha \tag{D.23}$$

where F_g is the magnitude of the force due to gravity, m is the mass of the object and g is the acceleration due to gravity.

Example: A 1 kg implant deflects $45^\circ = \pi/4$ radians on testing.

$$F_m = F_g \tan(\pi/4) = mg \tan(\pi/4) = F_g = 1\,kg \tag{D.24}$$

At a 45° deflection angle the implant magnetic force

is equal to and balanced by the implant gravitational force.

Drug Delivery Patch. See Transdermal Drug Patches.

Duty Cycle (f_d). The fraction of time the rf field is on during a pulse sequence.

(*MRSE*) The duty cycle f_d is

$$f_d = \frac{\sum_{i=1}^{n} \tau_{p_i}}{T_R} \tag{D.25}$$

for n pulses with pulse times τ_{p_i} and repetition time T_R.

E

E. Electric field. See Electric field.

$\mathbf{E}_1$. Electric field which is associated with applied rf field and the $\mathbf{B}_1$ field within the coil and bore. The $\mathbf{E}_1$ field is along the direction of $\mathbf{B}_0$ but near zero along the center of the bore and increases in magnitude radially toward the walls of the bore (Schaefer (2014)).

(*MRSE*) Following the discussion of Schaefer (Schaefer (2014)), an approximation to $\mathbf{E}_1$ can be obtained for an idealized birdcage coil using the assumption that the applied rf $\mathbf{B}_1$ field is homogeneous along the x-axis and therefore $\mathbf{B}_1 = (B_{1x} cos\omega t, 0, 0) = (B_1, 0, 0)$. Since by the geometry of the scanner and built in body coil, the rf coil conductors only lie along z, the current density only lies along z, and the magnetic vector potential is $\mathbf{A} = (A_x, A_y, A_z) = (0, 0, A_z)$. Therefore

$$\begin{aligned} \mathbf{B}_1 &= (B_1, 0, 0) = \nabla \times \mathbf{A} \\ &= \left(\frac{\partial A_z}{\partial y} - \frac{\partial A_y}{\partial z}, 0, 0\right) = \left(\frac{\partial A_z}{\partial y}, 0, 0\right) \end{aligned} \tag{E.1}$$

Then on integration $A_z = yB_1$. The electric field may be obtained from

$$\mathbf{E}_1 = -\frac{\partial \mathbf{A}}{\partial t} - \nabla V(\mathbf{r}) \tag{E.2}$$

under the assumption that $\nabla V(\mathbf{r}) = 0$ and timeharmonic vector potential of the form $\mathbf{A}(\mathbf{r},t) = \mathbf{A}(\mathbf{r})e^{i\omega t}$ then

$$\begin{aligned}\mathbf{E}_1 &= \left(0, 0, \frac{-\partial A_z}{\partial t}\right) = (0, 0, -i\omega A_z) \\ &= (0, 0, -i\omega y B_1) = (0, 0, E_z)\end{aligned} \tag{E.3}$$

so that

$$E_1 = E_z = \omega y B_1 \tag{E.4}$$

To summarize, while the rf $\mathbf{B}_1$ field is along the x-axis, the rf electric field in the coil is along the z-axis, like $\mathbf{B}_0$. The $\mathbf{B}_1$ field is homogeneous, while the rf electric field $\mathbf{E}_1$ is zero along the center of the coil and bore ($y = 0$ and therefore $E_1 = 0$). The electric field grows in magnitude moving from the center of the coil to its maximum at the periphery (Schaefer (2014)). For this reason, it is often advisable to place conductors such as leads down the center of the bore.

In addition, depending upon the field orientation, vertical or horizontal, the orientation of $\mathbf{E}_1$ with respect to an implant may be different and therefore the implant have different induced fields.

Ear Drum Rupture. In the event of a quench breach the boiled off cryogen may fill the room and raise the pressure to a level which could rupture ear drums (ACR Committee on MR Safety (2020)).

Ear Protection. Ear protection is used in MRI to protect the patient's hearing from damage due to the loud noise generated by the scanner. Ear protection such as earplugs are characterized by a noise reduction rating (NRR) (Occupational Safety and Health Administration Standard 29 CFR 1910.95). Ear plugs have noise reduction ratings on the order of 30 dB. See Noise Reduction Rating.

Earth's Magnetic Field. The magnetic field of the earth is approximately 0.5 G. In comparison a 1.5 T magnet corresponds to 15,000 G, or 30,000 times the Earth's magnetic field (Collins (2016)).

Echo Planar Imaging (EPI). A pulse sequence in which there is an excitation pulse such as a 90-degree pulse followed by the application of multiple gradients during which signal acquisition is performed (Ansorge and Graves (2016)).

Eddy Currents. Electrical currents induced in conductors via Faraday's Law (Fernow (2016)). These can cause heating or in the case of active implants, malfunction of the device.

(*MRSE*) Example: Suppose a time varying magnetic field $\mathbf{B}$ is applied perpendicular to the plane of a conducting ring of radius $r = 20\ cm$. The ring has a resistance of 2.5 Ohms. Find the voltage in the ring for a time varying field magnitude of $\partial B/\partial t = 20\ \mathrm{T}/s$. Since the electric field $\mathbf{E}$ is parallel to the differential element $d\mathbf{l}$ then by Faraday's Law

$$
\begin{aligned}
emf = El = \oint_C \mathbf{E} \cdot d\mathbf{l} &= -\frac{\partial}{\partial t} \int_S \mathbf{B} \cdot d\mathbf{A} \\
= -A\frac{\partial B}{\partial t} &= -\pi r^2 \frac{\partial B}{\partial t}
\end{aligned}
\tag{E.5}
$$

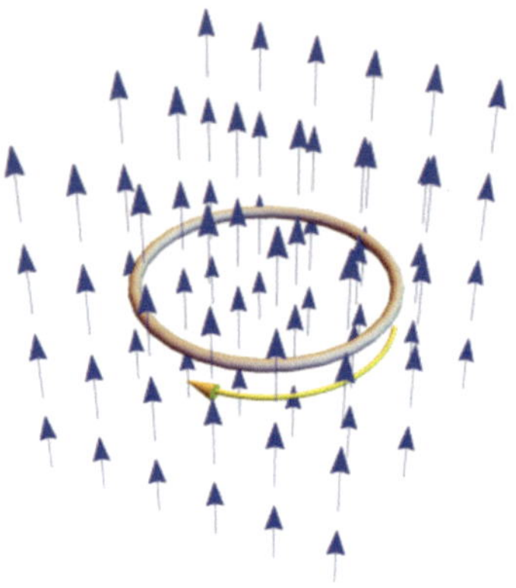

Figure E.1: Eddy current. Eddy current (yellow arrow) induced clockwise in a circular conductor by a magnetic flux density vector field (blue arrows) perpendicular to the plane of the conductor. The image is for a single point in time, as the field varies with time.

emf is the electromotive force in volts. The magnitude of the eddy current by Ohm's Law is

$$I = \frac{V}{R} = \frac{\pi (0.20\, m)^2\, 20\, \mathrm{T}/s}{2.5\Omega} = \frac{2.5V}{2.5\Omega} = 1A \quad \text{(E.6)}$$

Electric Dipole. Negative and positive charge separated by a distance.

(*MRSE*) An electric dipole moment $\boldsymbol{p}$ is defined as (Zangwill (2013))

$$\boldsymbol{p} = q\boldsymbol{d} \quad \text{(E.7)}$$

for positive and negative charges $+q$ and $-q$ separated by a distance vector $\boldsymbol{d}$.

Electric Dipole Field. A field from negative and positive charge separated by a distance.

(*MRSE*) In spherical coordinates the field from a dipole oriented along the z-axis is given by (Zangwill (2013))

$$\mathbf{E}_d(r,\theta) = \frac{p}{4\pi\varepsilon_0 r^3}(2\cos\theta\widehat{\mathbf{r}} + \sin\theta\widehat{\boldsymbol{\theta}}) \tag{E.8}$$

where p is the magnitude of the electric dipole moment, ε_0 is the permittivity of free space, and r is the distance from the dipole.

Electric Dipole Moment. See Electric Dipole.

Electric Field (E). A field which surrounds positive and negative charges. Electric fields are important in power deposition and heating in MRI.

(*MRSE*) The electric force per unit charge and is in the direction that a charge distribution would exert on a positive charge, which is a repulsive force for a positive charge distribution and attractive for a negative charge distribution on a positive charge (Zangwill (2013)).

For a point source at the origin

$$\mathbf{E}(\mathbf{r}) = \frac{1}{4\pi\varepsilon_0}\frac{q_0(0)}{|\mathbf{r}|^2}\frac{\mathbf{r}}{|\mathbf{r}|} = \frac{1}{4\pi\varepsilon_0}\frac{q_0(0)}{r^2}\widehat{\mathbf{r}} \tag{E.9}$$

where ε_0 is the permittivity of free space and $\mathbf{r}$ is the position vector relative to the charge at the origin, with unit vector directed between the charges of $\widehat{\mathbf{r}}$.

For a continuous charge distribution

$$\mathbf{E}(\mathbf{r}) = \frac{1}{4\pi\varepsilon_0}\int_V \frac{\rho_j(\mathbf{r}_j)}{|\mathbf{r}-\mathbf{r}_j|^2}\frac{\mathbf{r}-\mathbf{r}_j}{|\mathbf{r}-\mathbf{r}_j|}dV_j \qquad \text{(E.10)}$$

where the $\mathbf{r}_j$ represent the position of the source charge and the integral is taken over the volume V_j of the source charges.

Electric Permittivity of Free Space. See Permittivity of Free Space.

Electric Potential (V). Potential energy difference per unit charge. Also known as the electric scalar potential.

(*MRSE*) For a point source at the origin

$$V(\mathbf{r}) = \frac{1}{4\pi\varepsilon_0}\frac{q_0(0)}{|\mathbf{r}|} = \frac{1}{4\pi\varepsilon_0}\frac{q_0(0)}{r} \qquad \text{(E.11)}$$

where ε_0 is the permittivity of free space and $\mathbf{r}$ is the position vector relative to the charge at the origin.

For a continuous charge distribution

$$V(\mathbf{r}) = \frac{1}{4\pi\varepsilon_0}\int_V \frac{\rho_j(\mathbf{r}_j)}{|\mathbf{r}-\mathbf{r}_j|}dV_j \qquad \text{(E.12)}$$

where the $\mathbf{r}_j$ represent the position of the source charge and the integral is taken over the volume V_j of the source charges.

The electric field may be calculated from the magnetic vector potential $\mathbf{A}$ and electric scalar potential $V(\mathbf{r})$ by (Zangwill (2013))

$$\mathbf{E} = -\frac{\partial \mathbf{A}}{\partial t} - \nabla V(\mathbf{r}) \tag{E.13}$$

Electrical Safety. Equipment must be properly grounded. System covers should only be removed by qualified personnel. Liquids entering inside the scanner cover may cause short circuits. Breach of insulation or covers could have an associated risk of electric shock (Occupational Safety and Health Administration Standard 29 CFR 1910.301).

Electric Susceptibility (χ_E). A property of a substance which is a measure of its response to an external electric field and its associated polarization (Zangwill (2013)).

Electromagnet. A magnet formed by the creation of a magnetic field by a resistive electric current. Some MRI scanners generate their main magnetic fields through electromagnetism. Also known as a resistive magnet (Faulkner (2016), Halliday et al. (2014)).

Electromagnetic Interference (EMI). Term used to refer to the effects of electric and magnetic fields on electronic devices such as pacemakers (Indik et al. (2017)).

Electromagnetic Wave. A wave composed of orthogonal electrical and magnetic components which oscillate in time and move through space and can travel through vacuum and through media. In vacuum, electromagnetic waves travel at the speed of light. rf waves are electromagnetic waves.

(*MRSE*) Electromagnetic waves when far from the source can be modeled as uniform traveling plane waves. For a wave propagating along the z-axis as a function of time t, the electric component $\mathbf{E}(z,t)$ is along x and the magnetic component $\mathbf{B}(z,t)$ along y (Rao (2009), Griffiths (2017))

$$\mathbf{E}(z,t) = E_A \cos(kz - \omega t)\widehat{\mathbf{x}} = Re\left[E_A e^{i(kz-\omega t)}\right]\widehat{\mathbf{x}} \tag{E.14}$$

$$\mathbf{B}(z,t) = B_A \cos(kz - \omega t)\widehat{\mathbf{y}} = Re\left[B_A e^{i(kz-\omega t)}\right]\widehat{\mathbf{y}} \tag{E.15}$$

where E_A and B_A are the maximum amplitudes. The wavenumber $k = 2\pi/\lambda$ and Re denotes taking the real part of a complex function.

Electromotive Force (emf). (*MRSE*) The amount of work a device must do to move charge (Halliday et al. (2014)). SI unit is the volt (V).

Ellipsoid. A three dimensional shape which has two orthogonal dimensions which are the same, and one which is different. For example, an ellipsoid can be an elongated shape which is taller than wide in one dimension and smaller but the same in the other two dimensions (Cullity and Graham (2011), Schenck and Kangarlu (2014)).

Emergency Personnel. Personnel responding to an emergency in the MR environment. These may be hospital personnel such as a code team, or community personnel such as firemen, police, or paramedics.

It is desirable for all personnel including community emergency personnel to receive regular training in MR safety, although it must be acknowledged that response to emergencies in the MR environment is infrequent, and designated MR personnel should be available to help supervise and coordinate the response to an emergency in the MR environment to ensure the safety of all involved (ACR Committee on MR Safety (2020)).

EMI. See Electromagnetic Interference.

Entrapment. In the event of a quench breach in which cryogens are discharged into the room those inside the scanner room may be trapped from increased pressure against the main door making it impossible to open the door (ACR Committee on MR Safety (2020)).

Euler Rotation Matrix (R). (*MRSE*) When an interaction such as the magnetic susceptibility is known in one coordinate system, such as the principle axis system of the interaction in which the corresponding tensor is diagonal, but it is desired to know the interaction in another coordinate system, the interaction can be rotated into the new coordinate system. This is accomplished using the Euler rotation matrix given by (Arfken and Weber (2005), Mueller (2011))

$$\mathbf{R} = \begin{pmatrix} cos\alpha\, cos\beta\, cos\gamma - sin\alpha\, \sin\gamma & -cos\alpha\, cos\beta\, sin\gamma - sin\alpha\, cos\gamma & cos\alpha\, sin\beta \\ sin\alpha\, cos\beta\, cos\gamma + cos\alpha\, \sin\gamma & -sin\alpha\, cos\beta\, sin\gamma + cos\alpha\, cos\gamma & sin\alpha\, \sin\beta \\ -sin\beta\, cos\gamma & sin\beta\, sin\gamma & cos\beta \end{pmatrix} \tag{E.16}$$

with rotation angles α, β, and γ. For the example of rotating about the y-axis to an angle θ from the z-axis, $\beta = \theta$ and $\alpha = \gamma = 0$, the rotation matrix is given by

$$\mathbf{R}(\theta) = \begin{pmatrix} \cos\theta & 0 & \sin\theta \\ 0 & 1 & 0 \\ -\sin\theta & 0 & \cos\theta \end{pmatrix} \tag{E.17}$$

Another example used in other entries is a rotation about the y-axis from the x-axis to an angle θ with respect the z-axis, so that $\beta = \theta - \pi/2$ and $\alpha = \gamma = 0$ giving

$$\mathbf{R}(\theta) = \begin{pmatrix} \sin\theta & 0 & -\cos\theta \\ 0 & 1 & 0 \\ \cos\theta & 0 & \sin\theta \end{pmatrix} \tag{E.18}$$

Evaporation. A mechanism of cooling in MRI. Molecules near the surface of a liquid move into the gas phase taking away with them energy from the liquid resulting in evaporative cooling (Halliday et al. (2014)). Example: Sweating.

Expansion Ratio. (*MRSE*) The ratio of the volume of a cryogen in the gas state to its volume in the liquid state. For helium the expansion ratio is approximately 760 and for nitrogen it is approximately 705 (Faulkner (2016)). This can be determined as follows from the ideal gas equation (Halliday et al. (2014))

$$\mathrm{P}V_G = nRT \tag{E.19}$$

where P is pressure, V_G is volume of gas, n is moles, $R = 0.08206\,\text{L-}atm/\text{K-}mole$ is the gas constant, and T is temperature. Using the definition of molecular weight (M_W),

$$n = \frac{\rho V_L}{M_W} \tag{E.20}$$

where ρ is density and V_L is the liquid cryogen volume. The expansion ratio R_E may then be calculated

$$R_E = \frac{V_G}{V_L} = \frac{\rho RT}{M_W \text{P}} \tag{E.21}$$

For 1 L of liquid helium at 1 atmosphere pressure with a density of $0.125\,g/m\text{L}$ at 4.2 K, and molecular weight of $4.0\,g/mole$ (Lide (2004))

$$\begin{aligned} R_E = & \left(\frac{0.125\,g}{m\text{L}}\right)\left(\frac{1000\,m\text{L}}{1\,\text{L}}\right)\left(\frac{mole}{4.0\,g}\right) \\ & \cdot \left(\frac{0.08206\,\text{L-}atm}{\text{K-}mole}\right)\left(\frac{298\,\text{K}}{1}\right)\left(\frac{1}{1\,atm}\right) \\ & = 764 \end{aligned} \tag{E.22}$$

In the case of liquid nitrogen with a density of $0.807\,g/m\text{L}$ at 77 K with a molecular weight of $28.0\,g/mole$ (Lide (2004)), R_E works out to 705.

Example problem: Suppose there are 2000 L of liquid helium in a magnet which quenches and discharges the cryogen into the room. What is the volume V of helium gas? For the helium expansion ratio of 764

$$V = (764)(2000\,\mathrm{L}) = 1,528,000\,\mathrm{L} \qquad \text{(E.23)}$$

Exposure Limits. Various bodies such as the FDA, ICNIRP, OSHA, and IEC have established safety limits for exposures for various parameters associated with MRI. These may be different for different organizations. A summary is provided in tabular form below:

Table E.1: **FDA Static Magnetic Field Limits**

Age	*Static Field Limit (T)*
> 1 month	≤ 8
≤ 1 month	≤ 4

Reference: FDA (2014)

Table E.2: **FDA SAR Limits**

Anatomic Region	*SAR Limit W/kg*	*Time*
Whole Body SAR	≤ 4	15 min
Head SAR	≤ 3.2	10 min

Reference: FDA (2014)

Table E.3: **FDA Peripheral Nerve Stimulation Limits**

Anatomic Region	*PNS Limit*
Any	Patient Discomfort

Reference: FDA (2014)

Table E.4: **FDA Sound Pressure Limits**

Conditions	*Sound Pressure Level*
Peak Unweighted	140 dB
A-weighted rms With Hearing Protection	99 dB(A)

Reference: FDA (2014)

($MRSE$) Additional limits are given below, many of which depend on the frequency (ν) in Hz.

Table E.5: **IEC SAR Limits for Volume Transmit Coils**

Anatomic Region	*Normal Operating Mode W/kg*	*First Level Controlled Operating Mode W/kg*	*Second Level Controlled Operating Mode W/kg*	*Time*
Whole Body SAR	≤ 2	≤ 4	< IRB Limit	6 min
Head SAR	≤ 3.2	≤ 3.2	< IRB Limit	6 min
Partial Body SAR	$10 - 8\mathrm{R}_m$	$10 - 6\mathrm{R}_m$	< IRB Limit	6 min

Whole body SAR limits are derated by 0.25 W/kg for each degree C that the environmental temperature exceeds 25°C for the First Level Controlled Operating Mode. R_m = exposed mass/body mass. (Schaefer (2014))

Table E.6: **IEC SAR Limits for Local Transmit Coils**

Anatomic Region	*Normal Operating Mode W/kg*	*First Level Controlled Operating Mode W/kg*	*Second Level Controlled Operating Mode W/kg*	*Time*
Head or Trunk SAR over 10 g	≤ 10 W/kg	≤ 20	< IRB Limit	6 min
Extremity SAR over 10g	≤ 20	≤ 40	< IRB Limit	6 min

Reference: Schaefer (2014)

Table E.7: **IEC Peripheral Nerve Stimulation Limits**

PNS Limit	*Normal Operating Mode*	*First Level Controlled Operating Mode*	*Second Level Controlled Operating Mode*
Percent of Median Perception Threshold	80 %	100 %	< IRB Limit

Reference: McRobbie (2014)

Table E.8: **IEC Whole Body Temperature Limits**

Anatomic Region	*Normal Operating Mode °C*	*First Level Controlled Operating Mode °C*	*Second Level Controlled Operating Mode °C*
Whole Body	≤ 0.5	≤ 1.0	< IRB Limit

Reference: International Electrotechnical Commission (2010)

ICNIRP exposure limits are provided below. Note that the 2020 ICNIRP Guidelines For Limiting Exposure To Electromagnetic Fields (100 kHz to 300 GHz) replaces the 100 kHz to 300 GHz part of the 1998 ICNIRP radiofrequency guidelines, as well as the 100 kHz to 10 MHz part of the 2010 ICNIRP low-frequency guidelines.

Table E.9: **ICNIRP Static Magnetic Field Limits**

Population	*Anatomic Region*	*Static Field Limit* (T)
Occupational	Head and Trunk	≤ 2
Occupational	Limbs	≤ 8
General Population	Any Body Part	≤ 0.400

Note that for specific work applications exposures of the head and trunk up to 8 T may be justified, provided the environment is controlled and appropriate work practices are put in place to limit movement induced effects (International Commission on Non-Ionizing Radiation Protection (2009b)).

Table E.10: **ICNIRP Basic Restrictions for Time Varying Electric and Magnetic Fields Up to 10 GHz**

Frequency	*Current Density for Head and Trunk (mA/m (rms))*	*Whole Body Average SAR (W/kg)*	*Localized SAR Head and Trunk (W/kg)*	*Localized SAR Limbs (W/kg)*
		Occupational		
up to 1 Hz	40	-	-	-
1 - 4 Hz	$40/\nu$	-	-	-
4 Hz - 1 kHz	10	-	-	-
1 - 100 kHz	$\nu/100$	-	-	-
100 kHz - 10 MHz	$\nu/100$	0.4	10	20
10 MHz - 10 GHz	-	0.4	10	20
		General Public		
up to 1 Hz	8	-	-	-
1 - 4 Hz	$8/\nu$	-	-	-
4 Hz - 1 kHz	2	-	-	-
1 - 100 kHz	$\nu/500$	-	-	-
100 kHz - 10 MHz	$\nu/500$	0.08	2	4
10 MHz - 10 GHz	-	0.08	2	4

Note in particular the values for 10 MHz - 10 GHz. SAR is averaged over 6 minutes and localized SAR is taken over 10 g of tissue (International Commission on Non-Ionizing Radiation Protection (1998), International Commission on Non-Ionizing Radiation Protection (2009c)).

Table E.11: **ICNIRP Basic Restrictions for Electromagnetic Field Exposure 100 kHz to 300 GHz**

Frequency	*Whole Body Average SAR (W/kg)*	*Localized SAR Head/Torso (W/kg)*	*Local SAR Limbs (W/kg)*	*Local S_{ab} (W/m^2)*
		Occupational		
100 kHz - 6 GHz	0.4	10	20	NA
Greater than 6 - 300 GHz	0.4	NA	NA	100
		General Public		
100 kHz - 6 GHz	0.08	2	4	NA
Greater than 6 - 300 GHz	0.08	NA	NA	20

NA is "Not Applicable". Whole Body SAR is averaged over 30 minutes. Local SAR is averaged over 6 minutes and over a 10 g cubic mass of tissue. Local S_{ab} is averaged over 6 minutes and averaged over a 4 cm^2 square surface area of the body below 30 GHz. Above 30 GHz exposure is averaged over a 1 cm^2 square surface area of the body and is restricted to two times the 4 cm^2 restriction. (International Commission on Non-Ionizing Radiation Protection (2020)).

Table E.12: **ICNIRP Electric Field Limits Up to 10 MHz**

Anatomic Region	*Frequency*	*Internal Electric Field (V/m)*
	Occupational	
CNS tissues (head)	1 - 10 Hz	$0.5/\nu$
	10 - 25 Hz	0.05
	25 - 400 Hz	$2 \times 10^{-3}\nu$
	400 Hz - 3 kHz	0.8
	3 kHz - 10 MHz	$2.7 \times 10^{-4}\nu$
All tissues	1 Hz - 3 kHz	0.8
	3 kHz - 10 MHz	$2.7 \times 10^{-4}\nu$
	General Public	
CNS tissues (head)	1 - 10 Hz	$0.1/\nu$
	10 - 25 Hz	0.01
	25 Hz - 1 kHz	$4 \times 10^{-4}\nu$
	1 - 3 kHz	0.4
	3 - kHz - 10 MHz	$1.35 \times 10^{-4}\nu$
All tissues	1 Hz - 3 kHz	0.4
	3 kHz - 10 MHz	$1.35 \times 10^{-4}\nu$

Reference: International Commission on Non-Ionizing Radiation Protection (2010)

Table E.13: **Basic Restrictions for Electromagnetic Field Exposure 100 kHz to 10 MHz (Peak Spatial Values)**

Population	*Frequency*	*Induced Electric Field (V/m)*
Occupational	100 kHz - 10 MHz	$2.7 \times 10^{-4}\nu$
General Public	100 kHz - 10 MHz	$1.35 \times 10^{-4}\nu$

Applies to any region of the body and averaged as rms values over $2\,mm \times 2\,mm \times 2\,mm$ contiguous tissues (International Commission on Non-Ionizing Radiation Protection (2020))

Table E.14: **Reference Levels For Current Induced In Any Limb (100 kHz to 110 MHz)**

Population	*Frequency*	*Induced Electric Current I (mA)*
Occupational	100 kHz - 110 MHz	100
General Public	100 kHz - 110 MHz	45

Averaged over 6 min. (International Commission on Non-Ionizing Radiation Protection (2020))

Table E.15: **ICNIRP Reference Levels for Exposure to Time Varying Electric and Magnetic Fields Up to 10 MHz (Unperturbed *rms* Values)**

Frequency	*Electric Field Strength E (kV/m)*	*Magnetic Field Strength H* (A/m)	*Magnetic Flux Density B* (T)
		Occupational	
1 - 8 Hz	20	$1.63 \times 10^5/\nu^2$	$0.2/\nu^2$
8 - 25 Hz	20	$2 \times 10^4/\nu$	$2.5 \times 10^{-2}/\nu$
25 - 300 Hz	$500/\nu$	800	0.001
300 Hz - 3 kHz	$500/\nu$	$2.4 \times 10^5/\nu$	$0.3/\nu$
3 kHz - 10 MHz	0.17	80	1×10^{-4}
		General Public	
1 - 8 Hz	5	$3.2 \times 10^4/\nu^2$	$0.04/\nu^2$
8 - 25 Hz	5	$4 \times 10^3/\nu$	$0.005/\nu$
25 - 50 Hz	5	160	2×10^{-4}
50 Hz - 400 Hz	$250/\nu$	160	2×10^{-4}
400 Hz - 3 kHz	$250/\nu$	$6.4 \times 10^4/\nu$	$0.08/\nu$
3 kHz - 10 MHz	0.083	21	2.7×10^{-5}

Reference: International Commission on Non-Ionizing Radiation Protection (2010)

Table E.16: **ICNIRP Reference Levels for Exposure to Time Varying Electric and Magnetic Fields Up to 300 GHz (Unperturbed *rms* Values)**

Frequency	*Electric Field Strength E* (kV/m)	*Magnetic Field Strength H* (A/m)	*Magnetic Flux Density B* (μT)	*Equivalent Plane Wave Power Density S* (W/m^2)
Occupational				
Up to 1 Hz	-	1.63×10^5	2×10^5	-
1 - 8 Hz	2×10^4	$1.63 \times 10^5/\nu^2$	$2 \times 10^5/\nu^2$	-
8 - 25 Hz	2×10^4	$2 \times 10^4/\nu$	$2.5 \times 10^4/\nu$	-
0.025 - 0.82 kHz	$500/\nu$	$20/\nu$	$25/\nu$	-
0.82 - 65 kHz	610	24.4	30.7	-
0.065 - 1 MHz	610	$1.6/\nu$	$2.0/\nu$	-
1 - 10 MHz	$610/\nu$	$1.6/\nu$	$2.0/\nu$	-
10 - 400 MHz	61	0.16	0.2	10
400 - 2000 MHz	$3\sqrt{\nu}$	$0.008\sqrt{\nu}$	$0.01\sqrt{\nu}$	$\nu/40$
2 - 300 GHz	137	0.36	0.45	50
General Public				
Up to 1 Hz	-	3.2×10^4	4×10^4	-
1 - 8 Hz	10^4	$3.2 \times 10^4/\nu^2$	$4 \times 10^4/\nu^2$	-
8 - 25 Hz	10^4	$4,000/\nu$	$5,000/\nu$	-
0.025 - 0.8 kHz	$250/\nu$	$4/\nu$	$5/\nu$	-
0.8 - 3 kHz	$250/\nu$	5	6.25	-
3 - 150 kHz	87	5	6.25	-
0.15 - 1 MHz	87	$0.73/\nu$	$0.92/\nu$	-
1 - 10 MHz	$87/\sqrt{\nu}$	$0.73/\nu$	$0.92/\nu$	-
10 - 400 MHz	28	0.073	0.092	2
400 - 2,000 MHz	$1.375\sqrt{\nu}$	$0.0037\sqrt{\nu}$	$0.0046\sqrt{\nu}$	$\nu/200$
2 - 300 GHz	61	0.16	0.20	10

Note in particular the entries for 10 - 400 MHz (International Commission on Non-Ionizing Radiation Protection (1998), International Commission on Non-Ionizing Radiation Protection (2009c)).

Table E.17: **Reference Levels For Local Exposure To Electromagnetic Fields (100 kHz to 10 MHz)**

Population	*Frequency*	*Incident Electric Field E_{inc} (V/m)*	*Incident Magnetic Field H_{inc} (A/m)*
Occupational	100 kHz - 10 MHz	170	80
General Public	100 kHz - 10 MHz	83	20

Unperturbed rms values, for peak values (International Commission on Non-Ionizing Radiation Protection (2020))

Table E.18: **Reference Levels for Local Exposure to Electromagnetic Fields Up to 300 GHz Averaged Over 6 Minutes (Unperturbed *rms* Values)**

Frequency	*Incident Electric Field Strength* E_{inc} (V/m)	*Incident Magnetic Field* H_{inc} (A/m)	*Incident Power Density* S (W/m^2)
	Occupational		
0.1 - 30 MHz	$1504/\nu_M{}^{0.7}$	$10.8/\nu_M$	NA
> 30 - 400 MHz	139	0.36	50
> 400 - 2000 MHz	$10.58\nu_M{}^{0.43}$	$0.0274\nu_M{}^{0.43}$	$0.29\nu_M{}^{0.86}$
> 2 - 6 GHz	NA	NA	200
> 6 - < 300 GHz	NA	NA	$275/\nu_G{}^{0.177}$
300 GHz	NA	NA	300
	General Public		
0.1 - 30 MHz	$671/\nu_M{}^{0.7}$	$4.9/\nu_M$	NA
> 30 - 400 MHz	62	0.163	10
> 400 - 2000 MHz	$4.72\nu_M{}^{0.43}$	$0.0123\nu_M{}^{0.43}$	$0.058\nu_M{}^{0.86}$
> 2 - 6 GHz	NA	NA	40
> 6 - < 300 GHz	NA	NA	$55/\nu_G{}^{0.177}$
300 GHz	NA	NA	20

Note in particular the entries for 30 - 400 MHz. NA is "Not Applicable". ν_M and ν_G are frequency in MHz and GHz, respectively (International Commission on Non-Ionizing Radiation Protection (2020)).

Table E.19: **Reference Levels for Whole Body Exposure to Electromagnetic Fields Up to 300 GHz Averaged Over 30 Minutes (Unperturbed rms Values)**

Frequency	*Incident Electric Field Strength* E_{inc} (V/m)	*Incident Magnetic Field* H_{inc} (A/m)	*Incident Power Density* S (W/m^2)
Occupational			
0.1 - 30 MHz	$660/\nu_M{}^{0.7}$	$4.9/\nu_M$	NA
> 30 - 400 MHz	61	0.16	10
> 400 - 2000 MHz	$3\nu_M{}^{0.5}$	$0.008\nu_M{}^{0.5}$	$\nu_M/40$
> 2 - 300 GHz	NA	NA	50
General Public			
0.1 - 30 MHz	$300/\nu_M{}^{0.7}$	$2.2/\nu_M$	NA
> 30 - 400 MHz	27.7	0.073	2
> 400 - 2000 MHz	$1.375\nu_M{}^{0.5}$	$0.0037\nu_M{}^{0.5}$	$\nu_M/200$
> 2 - 300 GHz	NA	NA	10

Note in particular the entries for 30 - 400 MHz. NA is "Not Applicable". ν_M is frequency in MHz (International Commission on Non-Ionizing Radiation Protection (2020)).

Table E.20: **ICNIRP Motion Limits In a Static Magnetic Field and Exposure to Time Varying Fields of Less Than 1Hz**

Frequency f (Hz)	ΔB (T)	$B_{peak\,to\,peak}$ (T)	E (V/m)	E (V/m)	dB/dt (T/s)	dB/dt (T/s)
Critical Effect	Vertigo	Vertigo	PNS Effect	Phos-phenes	PNS Effect	Phos-phenes
Exposure Condi-tion	Uncon-trolled	Uncontrolled	Con-trolled	Con-trolled	Con-trolled	Uncon-trolled
0	2	-	-	-	-	-
0-1	-	2	-	-	-	-
0-0.66	-	-	1.1	1.1	2.7	2.7
0.66-1	-	-	1.1	$0.7/\nu$	2.7	$1.8/\nu$

ΔB is determined over 3 s periods. For controlled exposure a ΔB of 2 T may be exceeded (International Commission on Non-Ionizing Radiation Protection (2014)).

Table E.21: **ICNIRP Power Density Limits**

Population	*Frequency*	*Power Density*
Occupational	10 - 300 GHz	50 W/m^2
General Public	10 - 300 GHz	10 W/m^2

References: International Commission on Non-Ionizing Radiation Protection (1998) and International Commission on Non-Ionizing Radiation Protection (2009c)

Table E.22: **OSHA Power Density Limits**

Frequency	*Power Density*	*Energy Density*	*Time*
10 - 300 GHz	10 mW/cm^2	1 mW- hr/cm^2	0.1 hr

Reference: Occupational Safety and Health Administration Standard 29 CFR 1910.97

F

Faraday's Law. One of Maxwell's Equations stating that a magnetic field which changes with time is associated with a circulating electric field. This is a mechanism of heating of patients and objects as well as PNS and may also be a potential cause of implant malfunction.

(*MRSE*) Faraday's Law in differential and integral form are (Zangwill (2013))

$$\nabla \times \mathbf{E} = -\frac{\partial \mathbf{B}}{\partial t} \tag{F.1}$$

and

$$\oint_C \mathbf{E} \cdot d\mathbf{l} = -\frac{\partial}{\partial t} \int_S \mathbf{B} \cdot d\mathbf{A} \tag{F.2}$$

where $\mathbf{E}$ is the electric field and $\mathbf{B}$ is the magnetic flux density.

Simple example: Suppose a time varying field $\mathbf{B}$ of $\partial B/\partial t = 20$ T$/s$ is applied perpendicular to the plane of a conducting ring of radius $r = 20$ cm. What is the induced voltage?

Since the electric field $\mathbf{E}$ is parallel to the differential element $d\mathbf{l}$ then the electromotive force (emf) is

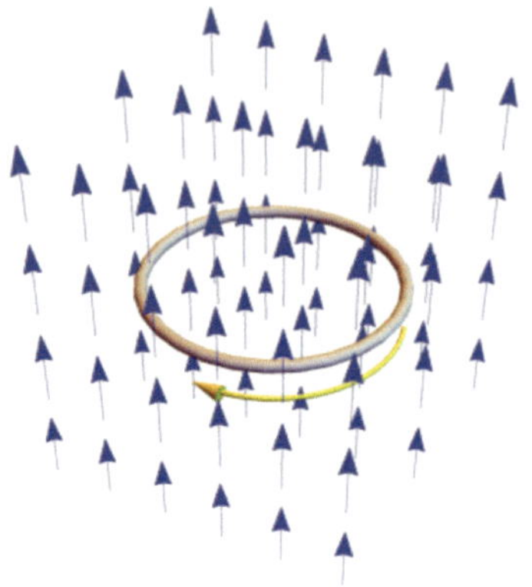

Figure F.1: Faraday's Law. Current (yellow arrow) induced clockwise in a circular conductor by a magnetic flux density vector field (blue arrows) perpendicular to the plane of the conductor. The image is for a single point in time, as the field varies with time.

$$\begin{aligned} emf = El = \oint_C \mathbf{E} \cdot d\mathbf{l} &= -\frac{\partial}{\partial t} \int_S \mathbf{B} \cdot d\mathbf{A} \\ &= -\frac{\partial}{\partial t} \int_S B dA \cos\theta = -A\frac{\partial B}{\partial t} = -\pi r^2 \frac{\partial B}{\partial t} \end{aligned} \tag{F.3}$$

Note the dependence on r^2, implying that there is a greater voltage induced for larger loops. This implies that larger patients will be at higher risk for heating. In addition, there is a dependence on the orientation of the loop with respect to the field with the maximum occurring for $\theta = 0$ when the magnetic field and plane of the loop are perpendicular.

Putting in the relevant values gives

$$emf = -\pi(0.20\,m)^2(20\,\mathrm{T}/s) = -2.5\,\frac{m^2\mathrm{T}}{s} = -2.5\,Volts \tag{F.4}$$

Fast Spin Echo (FSE). A type of pulse sequence involving an initial excitation pulse such as a 90-degree pulse followed by multiple acquisitions using multiple 180-degree pulses (Ansorge and Graves (2016), Brown et al. (2014)).

FDA. See Food and Drug Administration.

Ferromagnetic. Areas within a material have magnetic domains containing aligned atomic dipoles. When placed in a magnetic field, these domains become aligned, magnetizing the material, which persists to some extent when the field is removed. Ferromagnetic materials are characterized by magnetic susceptibility values on the order of 50 to 10,000. Example: Iron (Jiles (2015)). Ferromagnetic materials interact strongly with a magnetic field. See Magnetic Susceptibility.

Ferromagnetic Detector. Detection system which specifically detects ferromagnetic objects. Ferromagnetic detectors are recommended by the ACR Manual on MR Safety as part of the screening process while conventional metal detectors are not (ACR Committee on MR Safety (2020)).

Field of View (FOV). Physical dimensions corresponding to an MR image (Brown et al. (2014), Faulkner (2016)). In regard to MRI safety, field of view is one of the things which determines necessary gradient strength and sampling rate to generate an image.

(*MRSE*) The field of view is related to the bandwidth $\Delta\omega_i$ for i equal pe (phase encode) and fe (frequency encode) directions by (Brown et al. (2014))

$$FOV_{pe} = \frac{\Delta\omega_{pe}}{\gamma G_{pe}} \tag{F.5}$$

$$FOV_{fe} = \frac{\Delta\omega_{fe}}{\gamma G_{fe}} \tag{F.6}$$

with associated gradient G_i and bandwidth $\Delta\omega_i$. Note that for the same bandwidth a smaller field of view requires a larger gradient, which produces a higher dB/dt with greater potential for induced voltages.

Field Plot. A plot of various types of field parameters in MRI relative to the physical dimensions of the scanner. These may be isogauss lines although plots with lines of spatial gradient or force product are also common with respect to the static field. Plots of magnetic field or relative magnetic field generated by coils are also often provided by manufacturers. The values of the various field parameters at points in the scanner and scanner room is important for determining the safety of equipment, patients, and personnel, particularly patients with implants.

Static Field Plot:

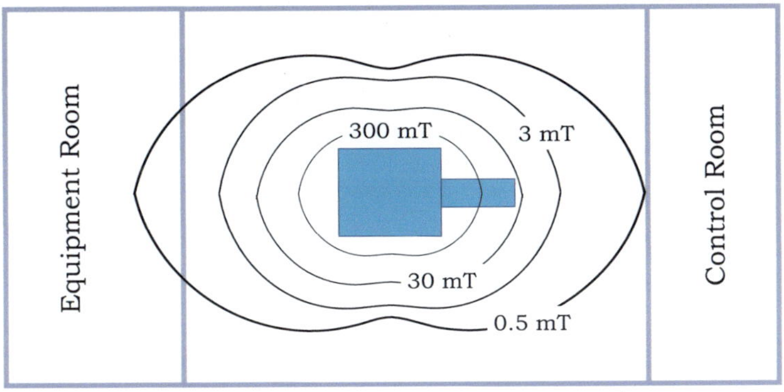

Figure F.2: Static Magnetic Field Map. Lines of constant field (isogauss lines) are shown around the scanner. Note that the figure is for a hypothetical scanner using an approximate field function and not necessarily to scale.

Example: The 5 G line extends into the equipment room (T/F)? True. 5 G = 0.5 mT, so from Figure F.2 this does encroach on the equipment room but not the control room. A person standing at the end of the table would experience a field of around 5 G (T/F)? False. From Figure F.2 the field at the end of the table is approximately 30 mT = 300 G.

Spatial Gradient Plot:

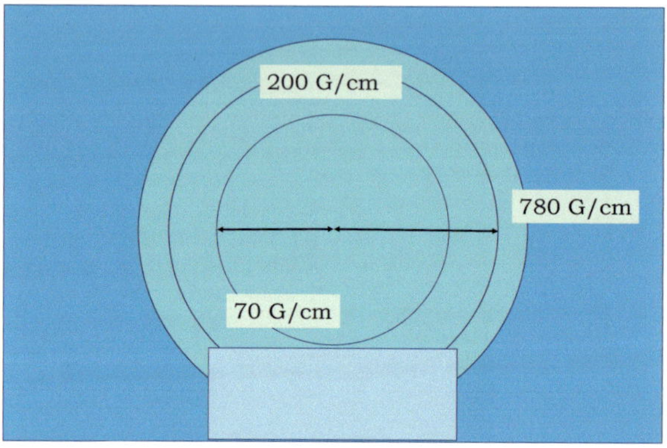

Figure F.3: Spatial Gradient Field Map. View down the table and bore of a scanner with spatial gradient contour lines. Note that the figure is for a hypothetical scanner using an approximate field function and not necessarily to scale.

Example: A scanner has the axial spatial gradient plot shown in Figure F.3. An implant centered in the patient has labeling indicating recommended use with a maximum spatial gradient less than 780 G/cm. The implant is safe to scan. True. The patient's implant is unlikely to contact a spatial gradient more than 780 G/cm in the center of the patient.

Transmit Coil Field Plot:

The scanner operating manual will usually provide a plot of the ratio of the square of B_1 at a position along z from the isocenter with respect to the the square of B_1 at the isocenter, for example:

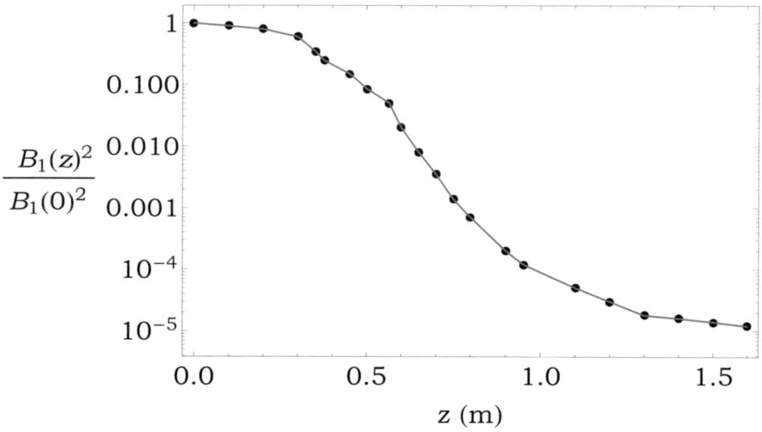

Figure F.4: Relative rf power distribution.

Note that Figure F.4 is a logarithmic plot. Starting at approximately 50 cm from the isocenter along z, the square of the field, which is proportional to SAR, is 10% of the full rf found at isocenter, but by 80 cm it is at 1/1000 of the full value.

Example: What percent of the square of the maximal field would a person standing 1 m from isocenter receive? From Figure F.4 they would receive 0.0001 or 0.01%.

Time Varying Gradient Field Plot:

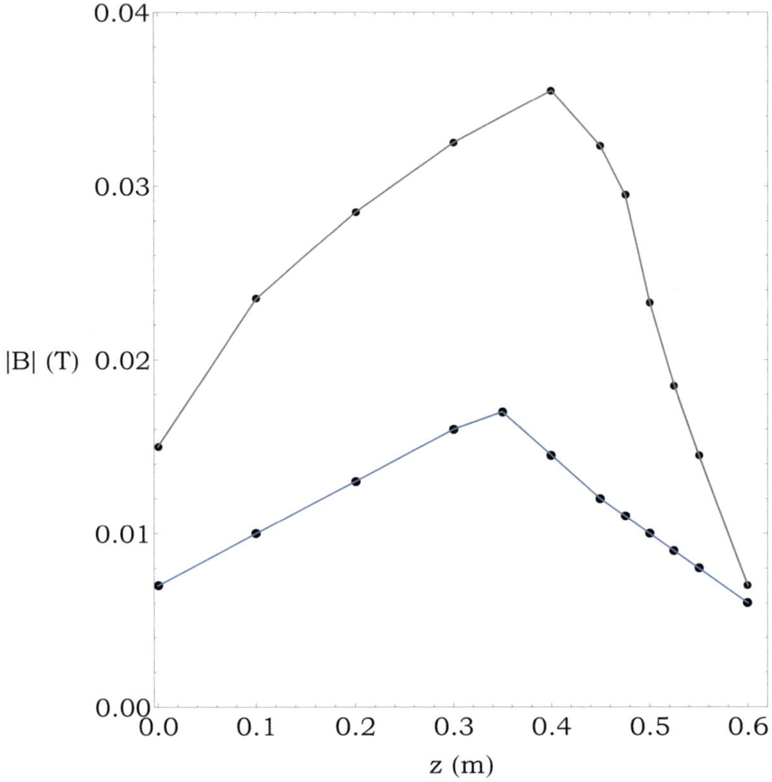

Figure F.5: Spatial distribution of the magnitude of the gradient field. The lower blue line represents values near the patient and farther from the periphery of the bore, while the upper line represents the field close to the periphery of the bore.

Example: The maximum time varying gradient field occurs approximately 40 cm from isocenter (T/F)? True. Looking at the upper curve in Figure F.5, the maximum occurs near 0.4 m = 40 cm. This would be the area of the highest dB/dt from the time varying gradient coils with most risk for PNS.

Firearms. Firearms may be attracted toward the magnet, and the firing pin and safety could move and discharge the weapon (Delfino et al. (2019)). Personnel carrying firearms such as police and hospital security. Firearms or ferromagnetic weapons of any type should not be permitted in Zone III unless deemed absolutely essential for maintaining security. Anyone carrying a firearm should be instructed to refrain from entering Zone IV (ACR Committee on MR Safety (2020)). If a firearm becomes stuck to the magnet, it should be unloaded by an individual familiar with its operation prior to its being removed from the magnet (Kanal and Shaibani (1994)).

First Level Controlled Operating Mode. Mode of MRI scanner operation in which the output may cause physiological stress. Quantitative limits for this mode have been published by the IEC (Keevil (2014), Schaefer (2014)). See Exposure Limits.

Flip Angle (α_p). When the $\mathbf{B}_1$ field is turned on for a period time, the flip angle is the angle through which the nuclear magnetization travels. For spin echo sequences these are typically 90-degrees and 180-degrees, while gradient echo sequences typically use less than 90-degree pulses to excite the magnetization (Ansorge and Graves (2016)). The SAR depends upon the square of the flip angle.

(*MRSE*) The flip angle α_p is given by (Brown et al. (2014))

$$\alpha_p = \gamma B_1 \tau_p \tag{F.7}$$

where γ is the gyromagnetic ratio, B_1 is the magnitude of the applied rf field, and τ_p is the time the pulse is on (pulse length).

Flow Potential. A voltage generated perpendicular to the flow of blood related to a Lorentz force induced charge separation within the vessel (Schenck and Kangarlu (2014)). Responsible for T wave distortion on EKG.

(*MRSE*) The flow potential arises in the setting of the magnetohydrodynamic effect and is given by

$$V_f = \int \mathbf{v} \times \mathbf{B}_0 \cdot d\mathbf{l} \tag{F.8}$$

where V_f is the flow potential from the velocity of flowing blood $\mathbf{v}$ moving through the static field $\mathbf{B}_0$. Note that the potential will be maximized when the direction of flow and $\mathbf{B}_0$ are at right angles, such as would occur for blood flowing across the top of the aortic arch.

Flux Lines. Visual model of a vector field using straight or curved lines with arrows along the lines (Rao (2009)).

Fog. In the event of a quench breach, the cold cryogens may produce a fog (ACR Committee on MR Safety (2020)).

Food and Drug Administration (FDA). Relevant to MRI safety, the mission of the FDA is to protect public health by ensuring the safety, efficacy, and security of drugs and medical devices (Kainz (2014)).

Force Due to Gravity ($\mathbf{F}_g$). The force that gravity exerts on an object, which is the weight of the object (Halliday et al. (2014)). The force due to gravity is important in MRI safety since it is generally assumed that the force due to gravity is safe for an implant and consequently any force exerted by the magnet which is less than the force of gravity will be safe.

(*MRSE*) Common weight measures are the pound and the Newton. The force due to gravity is determined by (Halliday et al. (2014))

$$\mathbf{F}_g = -m\mathbf{g} \tag{F.9}$$

or in terms of the corresponding magnitudes

$$F_g = mg \tag{F.10}$$

where F_g is the force due to gravity, m is mass, and g is the acceleration due to gravity.

Force Product. The product of the magnitude of the field and the magnitude of the spatial gradient. In terms of MRI safety the displacement force experienced by an unsaturated magnetic object is proportional to the force product. The maximum force product and hence the maximum displacement force is typically found near the face of the bore. Along the long axis of the table near the scanner entrance this may be approximated as

$B_0 dB_0/dz$ (Schenck (2000), Schenck and Kangarlu (2014), ASTM F2052-15 (2015)). The units are T^2/m.

(*MRSE*) More generally, the magnitude of the force product F_p is given by (Schenck (2000), Schenck and Kangarlu (2014), ASTM F2052-15 (2015))

$$F_p = B_0 \left|\nabla B_0\right| \tag{F.11}$$

which for $\mathbf{B}_0 = \big(0, 0, B_0(z)\big) = \big(0, 0, B_0\big)$ collapses to $B_0 \dfrac{B_0}{dz}$. See Magnetic Displacement Force.

Example: A Conditional implant specifies 1.5 T and a maximum spatial gradient of 720 G/cm. For a magnet with the corresponding parameters 1.5 T and 790 G/cm, how does this compare? Considering the magnitude of the magnetic displacement forces (F_m) in the two cases (ASTM F2052-15 (2015)) assuming an unsaturated object

$$\frac{F_{m_2}}{F_{m_1}} = \frac{F_{p_2}}{F_{p_1}} = \frac{B_{0_2}\left|\nabla B_0\right|_2}{B_{0_1}\left|\nabla B_0\right|_1} = \frac{1.5\,\mathrm{T}\,(790\,G/cm)}{1.5\,\mathrm{T}\,(720\,G/cm)} = 1.1 \tag{F.12}$$

If we assume that the conditional parameters would exert at most the equivalent to the force of gravity, then the scanner is expected to exert at most on the order of 1.1 times the force of gravity on the implant in terms of the displacement force. Note that the location of the maximum spatial gradient and the maximum force product may not be the same, and the maximum force product may be less than the product of the nominal field strength and maximum spatial gradient, since the maximum spatial gradient and location of the nominal field strength may not be coincident.

Foreign Bodies. Foreign bodies which are not medical implants, for example bullets or shrapnel. All patients with a history of orbital trauma for which they sought medical attention should be cleared by plain radiography or review of prior contiguous slice CT or MRI obtained since the injury (ACR Committee on MR Safety (2020)). The final determination whether or not to scan any foreign body rests with the Level 2 attending MR radiologist, the MR medical director, or designated Level 2 personnel using predetermined protocols. Assessment of risks includes consideration of the likelihood of and effects of forces on and currents in the foreign object which might occur during scanning. This includes an assessment of the relative location to vital structures and size (ACR Committee on MR Safety (2020)).

FOV. Field of view.

Frequency (ν). In an electromagnetic wave the frequency is how fast the wave cycles in time (Halliday et al. (2014)). For rf the frequencies are on the order of megahertz (MHz). For protons at 1.5 T the frequency is approximately 64 MHz.

Frequency Encoding. Digitization of the oscillating MRI signal during the detection period of a pulse sequence in the presence of an applied field gradient (Brown et al. (2014))

Frostbite. A low temperature thermal injury resulting in tissue damage. Relevant to MRI safety this can occur if skin comes into contact with the magnet's cryogens, for example from a quench breach (ACR Committee on MR Safety (2020)). First aid includes carefully removing clothing without damaging underlying skin and flushing the affected area with lukewarm water.

Full Stop and Final Check. A "full stop and final check" by the MRI technologist is recommended prior to taking the patient from Zone III to Zone IV to verify patient identification, screening has been appropriately performed, and there has been no change in patient or equipment status. This is analogous to the"timeout" procedure required by TJC (Greenberg et al. (2019), ACR Committee on MR Safety (2020)).

G

g. The acceleration due to gravity, $g = 9.8\,m/s^2$ (Halliday et al. (2014)).

Gadolinium. Lanthanide series rare earth metal. The Gd^{3+} ion is chelated by an organic molecule to form a contrast agent. It has seven unpaired electrons making it a powerful paramagnetic relaxation agent (Rogowska et al. (2018)).

Gadolinium Based Contrast Agents. Drugs used in MRI to increase contrast conspicuity between tissues and between pathological lesions and adjacent tissues (ACR Committee on Drugs and Contrast Media (2020)). These agents consist of a gadolinium ion bound to an organic molecule. The ion is referred to as being chelated to the organic molecule which serves as the ligand. Contrast agents may be categorized for the purposes of MRI safety with respect to their chemical structures in terms of how the ion is bound to the chelate as ionic (I) vs. nonionic (N) and structure of the complex as linear (L) vs. macrocyclic (M). Some contrast agents can also bind to protein.

In the United States the following contrast agents, with manufacturer given in brackets, are of importance:

Gadoterate meglumine (Dotarem)-(M,I) [Guerbet], Gadoterate meglumine (Clariscan)-(M,I) [GE Healthcare], Gadoxetate disodium (Eovist)-(L,I) [Bayer Healthcare], Gadobutrol (Gadavist)-(M,N) [Bayer Healthcare], Gadopentetate dimeglumine (Magnevist)-(L,I) [Bayer Healthcare], Gadobenate dimeglumine

(Multihance)-(L,I) [Bracco], Gadodiamide (Omniscan)-(L,N) [GE Healthcare], Gadoversatamide (Optimark)-(L,N) [Mallinckrodt], and Gadoteridol (Prohance)-(M,N) [Bracco]. Gadoterate meglumine is sold under different two different names by two different companies.

Gadoversatamide (Optimark) was withdrawn from the US market in 2018 (McDonald et al. (2018)). Gadopentetate dimeglumine (Magnevist) was also withdrawn from the US market in 2019.

In terms of stability ionic bonds are more stable than nonionic, and macrocyclic complexes are more stable than linear agents. It is thought that the more stable the chelate the safer it is in terms of avoiding the effects of free gadolinium. Note that Gadoterate meglumine (Dotarem and Clariscan) which is an ionic macrocyclic is expected to be the most stable while Gadodiamide (Omniscan)and Gadoversatamide (Optimark) are linear nonionic, and expected to have the least stability. In fact agents such as Gadodiamide (Omniscan) have the most NSF cases associated with them (ACR Committee on Drugs and Contrast Media (2020)).

In terms of risk stratification for NSF, gadolinium agents are characterized into three groups with Group I having the most risk:

Group I agents: Gadodiamide (Omniscan), Gadoversatamide (Optimark), and Gadopentetate dimeglumine (Magnevist).

Group II agents: Gadobenate dimeglumine (Multihance), Gadoteridol (Prohance), Gadoterate meglumine (Dotarem and Clariscan), and Gadobutrol (Gadavist).

Group III agents: Gadoxetate disodium (Eovist)

Gadolinium Contrast Group I Agents. Gadodiamide (Omniscan), Gadoversatamide (Optimark), and Gadopentetate dimeglumine (Magnevist) (ACR Committee on Drugs and Contrast Media (2020)).

Gadolinium Contrast Group II Agents. Gadobenate dimeglumine (Multihance), Gadoteridol (Prohance), Gadoterate meglumine (Dotarem and Clariscan), and Gado-butrol (Gadavist) (ACR Committee on Drugs and Contrast Media (2020)).

Gadolinium Contrast Group III Agents. Gadoxetate disodium (Eovist) (ACR Committee on Drugs and Contrast Media (2020)).

Gadolinium Contrast in Pregnancy. In general, administration of gadolinium in pregnancy should be avoided, as gadolinium based contrast agents cross the placenta and are excreted into the amniotic fluid, where they may remain for some time. Since the associated risk to the fetus is not definitely known and could be harmful, the decision to administer a gadolinium based contrast agent is to be made by the Level 2 attending radiologist after an appropriate risk/benefit assessment. After discussion with the referring physician the radiologist should document that the imaging information to be obtained from MRI cannot be obtained without contrast or from another modality such as sonography, the information obtained will affect care during pregnancy, and it is not reasonable to wait until after delivery. Written consent from the patient should be obtained (ACR Committee on Drugs and Contrast Media (2020)).

Gadolinium Contrast Reaction. Adverse physiological response to gadolinium based contrast media administration (ACR Committee on Drugs and Contrast Media (2020)). These may be divided into allergic and nonallergic reactions. Gadolinium reactions are quoted to occur 0.07 to 2.4% of the time, with allergic type reactions occurring 0.004% to 0.7% of the time. Severe allergic anaphylactic reactions occur with a 0.001% to 0.01% incidence. Deaths from gadolinium contrast administration, although extremely rare, have been reported.

A prior reaction to a gadolinium based contrast agent increases the chance of subsequent reactions by a factor of eight. Gadobenate dimeglumine (Multihance) is contraindicated in its FDA labeling in the presence of prior history of gadolinium reaction.

Patients with asthma and allergies are also more likely to react to gadolinium.

Per the 2018 ACR Contrast Manual there is no cross reactivity between iodinated contrast and gadolinium based contrast.

Treatment of acute allergic gadolinium reactions follows the same protocol as for iodinated agents. For mild reactions expectant management may be sufficient, and the reaction may abate on its own. The mainstays of pharmacological intervention are diphenhydramine and epinephrine (ACR Committee on Drugs and Contrast Media (2020)). See Treatment of Gadolinium Contrast Reactions.

Because of the chemical nature of gadolinium and small volumes injected, significant adverse effects from extravasation are considered extremely unlikely (ACR Committee on Drugs and Contrast Media (2020)).

Gadolinium Deposition Disease (GDD). Putative disease process related to gadolinium in which patients with normal renal function exposed to gadolinium have a number of vague symptoms following administration of gadolinium (Semelka et al. (2016)).

Gadolinium Ionic Agents. Gadoterate meglumine (Dotarem and Clariscan), Gadoxetate disodium (Eovist), Gadopentetate dimeglumine (Magnevist), and Gadobenate dimeglumine (Multihance) (ACR Committee on Drugs and Contrast Media (2020)).

Gadolinium Linear Agents. Gadoxetate disodium (Eovist), Gadopentetate dimeglumine (Magnevist), Gadobenate dimeglumine (Multihance), Gadodiamide (Omniscan), and Gadoversatamide (Optimark) (ACR Committee on Drugs and Contrast Media (2020)).

Gadolinium Linear Ionic Agents. Gadoxetate (Eovist), Gadopentetate dimeglumine (Magnevist), and Gadobenate (Multihance) (ACR Committee on Drugs and Contrast Media (2020)).

Gadolinium Linear Nonionic Agents. Gadodiamide (Omniscan), and Gadoversatamide (Optimark) (ACR Committee on Drugs and Contrast Media (2020)).

Gadolinium Macrocyclic Agents. Gadobutrol (Gadavist), Gadoterate meglumine (Dotarem and

Clariscan), and Gadoteridol (Prohance) (ACR Committee on Drugs and Contrast Media (2020)).

Gadolinium Macrocyclic Ionic Agents. Gadoterate meglumine (Dotarem and Clariscan) (ACR Committee on Drugs and Contrast Media (2020)).

Gadolinium Macrocyclic Nonionic Agents. Gadobutrol (Gadavist), and Gadoteridol (Prohance) (ACR Committee on Drugs and Contrast Media (2020)).

Gadolinium Nonionic Agents. Gadoteridol (Prohance), Gadodiamide (Omniscan), Gadoversatamide (Optimark), and Gadobutrol (Gadavist) (ACR Committee on Drugs and Contrast Media (2020)).

Gadolinium Protein Binding Agents. Gadoxetate disodium (Eovist), and Gadobenate dimeglumine (Multihance) (McDonald et al. (2018)).

Gadolinium Storage Condition. Condition of gadolinium being stored in the body after contrast administration. This is associated with the less stable agents. The pathophysiological significance is incompletely understood. Gadolinium has been found in patients with normal renal function in organs such as the brain and skeleton (Semelka et al. (2016)).

Gadolinium Toxicity. Free gadolinium ion is toxic through its competition with calcium ion metabolism. The chelated form in gadolinium contrast agents is much less toxic; however, there is toxicity such as nephrogenic systemic fibrosis (NSF) and perhaps gadolinium deposition disease (GDD) and may be related to small amounts of free gadolinium dissociated from the chelate (Rogowska et al. (2018)).

Gauss (G). Unit of magnetic flux density in the cgs system. Example: Earth's magnetic field is 0.5 G. 1 T = 10,000 G.

Gauss's Law for Electricity. (*MRSE*) One of Maxwell's Equations describing how the total electric flux through a closed surface is related to the charged enclosed inside the enclosing surface. Gauss's Law for Electricity in differential and integral form are (Zangwill (2013))

$$\nabla \cdot \mathbf{E} = \frac{\rho_q}{\varepsilon_0} \tag{G.1}$$

and

$$\oint_S \mathbf{E} \cdot d\mathbf{A} = \frac{Q_e}{\varepsilon_0} \tag{G.2}$$

$\mathbf{E}$ is the electric field, Q_e is the charge enclosed within a surface, ρ_q is the volume charge density of the object from which the electric field is emanating, and $d\mathbf{A}$ is the differential surface element normal to the surface.

Example: Find the surface charge density for a perfectly conducting hollow sphere of radius R_s with an electric field $\mathbf{E}$. Since the electric field radiates outward at every point, the electric field vector and area vector are parallel at every point and the angle θ between them is zero so that

$$\oint_S \mathbf{E} \cdot d\mathbf{A} = \oint_S EA\cos\theta = EA = \frac{Q_e}{\varepsilon_0} \tag{G.3}$$

where A is the area of the sphere. Since for a perfect conductor the charge resides on the surface of the sphere, and

$$\frac{Q_e}{\varepsilon_0} = \frac{\sigma_q A}{\varepsilon_0} \tag{G.4}$$

where σ_q is the surface charge density, then

$$\sigma_q = \varepsilon_0 E \tag{G.5}$$

Gauss's Law for Magnetism. (*MRSE*) One of Maxwell's Laws stating that the magnetic flux density $\mathbf{B}$ always curves back on itself. That is to say that the magnetic flux lines always run between the north and south poles of the magnetic and never out into space. This implies there are no magnetic monopoles. Gauss's Law for Magnetism in differential and integral form are (Zangwill (2013))

$$\nabla \cdot \mathbf{B} = 0 \tag{G.6}$$

and

$$\oint_S \mathbf{B} \cdot d\mathbf{A} = 0 \tag{G.7}$$

GBCM. See Gadolinium Based Contrast Media.

General Anesthesia. Sedation level in which the patient is unarousable even with painful stimuli. Airway protection and ventilatory function are usually impaired and cardiovascular function may be impaired (Reinking (2014)). Some MRI patients will require general anesthesia.

GFR. See Glomerular Filtration Rate.

Glomerular Filtration Rate (GFR). A measure of the ability of the kidneys to filter the blood representing the amount of substance cleared from the blood in a given period of time. This is an important parameter in the risk assessment for the administration of intravenous contrast. The glomerular filtration rate (GFR) can be measured directly, for example using radioisotope measurements, but these methods are time consuming and expensive. Therefore, estimated GFR (eGFR) is obtained as a calculated value from equations such as the Modification of Diet in Renal Disease (MDRD) equation in adults and the Bedside Schwartz Equation in children. Units of eGFR are $mL/min/1.73\,m^2$ (Simon et al. (2011), Parsh et al. (2015), ACR Committee on Drugs and Contrast Media (2020), ACR Committee on MR Safety (2020)).

(*MRMD*) The MDRD Equation is (Parsh et al. (2015))

$$\text{eGFR} = 175(S_{Cr}^{-1.154})(Age^{-0.203})(f)(a) \qquad \text{(G.8)}$$

where S_{Cr} is serum creatinine in mg/dL, Age is in years, and $f = 0.742$ for females and $f = 1$ for males. $a = 1.212$ for African Americans and $a = 1$

otherwise. There is concern that the MDRD Equation can incorrectly diagnose elderly white females to have a low eGFR.

The Bedside Schwartz Equation is (ACR Committee on Drugs and Contrast Media (2020))

$$\text{eGFR} = \frac{0.41(Height)}{S_{Cr}} \tag{G.9}$$

where $Height$ is height in cm and S_{Cr} is serum creatinine in mg/dL.

Golay Coil. A set of four saddle coils used to produce time varying transverse gradients in the x and y directions perpendicular to the main magnetic field (Brown et al. (2014)).

Gradient. A term referring to how the value of something changes with respect to another parameter. In MRI this is the strength of the magnetic field with distance. There are two types of gradients. There is the static field gradient which is how fast the main magnetic field falls off moving away from the magnet, for example dB_0/dz. This magnetic field changes with distance but not with time, hence the name static. The other gradient is the time varying gradients where gradients of, for example, dB_z/dx are turned on and off to make the MR image, so that they have an associated dB/dt. Since this turns on and off with time it is time varying but it also changes with distance over the field of view when it is applied (Brown et al. (2014)).

It is important not to confuse the two! The spatial gradient of the static field is on the order of 10 T/m, while the applied time varying gradients are on the order of 40 mT/m.

A gradient can be expressed in one dimension (for example the z-axis) as dB_0/dz. In three dimensions the gradient is expressed by a partial differential vector operator (Griffiths (2017)).

The time varying gradient strength needed to achieve a given field of view depends on the bandwidth of relevance. See Bandwidth.

Gradient Coils. Coils used to generate the time varying gradients used to generate the images. These use Golay and Maxwell coils. The gradient coils have a spatial profile such that the gradient field is zero at isocenter and maximal near the ends of the coils near the ends of the bore (Brown et al. (2014)).

Gradient Echo (GRE). A form of MRI echo in which a pair of gradients of the same strength but different signs are applied. Gradient echoes are typically lower SAR than a spin echo (Ansorge and Graves (2016), Brown et al. (2014)).

Gyromagnetic Ratio (γ). A constant related to a given atomic nucleus. Clinical MRI most often uses the proton as the imaged nucleus, which has a gyromagnetic ratio $\gamma = 42.6$ MHz/T (Faulkner (2016), Brown et al. (2014)). The gyromagnetic ratio determines the resonance frequency of a nucleus for a given field strength by $\omega = \gamma B$, where ω is angular frequency and B is the magnitude of the magnetic flux density (Brown et al. (2014)).

H

H. Magnetic field.

(*MRSE*) The relationship of the magnetic field **H** to the magnetic flux density **B** in a vacuum is (Zangwill (2013))

$$\mathbf{B} = \mu_0 \mathbf{H} \tag{H.1}$$

where μ_0 is the magnetic permeability of free space.

With the introduction of matter this becomes

$$\mathbf{B} = \mu_0 (\mathbf{H} + \mathbf{M}) \tag{H.2}$$

where **M** is the magnetization induced in the matter by **H**. There is also a relationship between **H** and **M**

$$\mathbf{M} = \boldsymbol{\chi} \mathbf{H} \tag{H.3}$$

where $\boldsymbol{\chi}$ is the magnetic susceptibility tensor. See **B**.

$\mathbf{H}_0$. Magnetic field of the MRI scanner.

(*MRSE*) The relationship of the scanner magnetic field $\mathbf{H}_0$ to the magnetic flux density associated with the scanner $\mathbf{B}_0$ in a vacuum is (Zangwill (2013))

$$\mathbf{B}_0 = \mu_0 \mathbf{H}_0 \tag{H.4}$$

where μ_0 is the magnetic permeability of free space.

With the introduction of matter this becomes

$$\mathbf{B}_0 = \mu_0(\mathbf{H}_0 + \mathbf{M}) \tag{H.5}$$

where **M** is the magnetization induced in the matter by $\mathbf{H}_0$. See **H**.

Heart Rhythm Society 2017 Consensus Guidelines. A set of guidelines published by the Heart Rhythm Society regarding the effects of scanning CIEDs and the procedures for scanning patients with CIEDs (Indik et al. (2017)).

Heating. In the context of MRI safety, heating can arise in multiple settings and from multiple mechanisms. Global heating of the patient can occur in the setting of high SAR or SED. This is related to generalized inductive heating via Faraday's Law from the time alternating magnetic component of the rf electromagnetic field. This is particularly true of patients with impaired thermoregulation including obese patients. Larger patients also will intrinsically be at increased risk for heating through Faraday's Law. Furthermore, local hot spots may develop within the patient due to inhomogeneous SAR (Schaefer (2014)).

Near field effects next to the rf coil may also cause heating.

If portions of the patient form a large loop particularly with sharp contact points, such as the mid thighs touching in a small focal area, then eddy currents within the loop can lead to heating (Mandel et al. (2017)).

In the context of implants or external conductors

such as monitoring leads, jewelry, or tattoos, currents can similarly form secondary to magnetic induction leading to burns. In addition, in the case of implants and other conductors, the electric field component of the rf electromagnetic wave can produce currents in conductors leading to heating. Note that in the case of conductors, the object itself may not heat significantly, but the surrounding tissue will (Nyenhuis (2014), Nyenhuis (2018)).

The degree of heating in any situation may be mitigated by an individual patient's adaptive response to heating such as blood flow regulation (Schaefer (2014), Shrivastava (2018)).

A passive device with deployed dimensions of less than 2 cm is expected to generate less than a 2 °C temperature rise over one hour of rf exposure at at 1.5 T or 3 T, provided that multiple replicas of the device are not placed within 3 cm of each other to avoid rf coupling between the devices (ASTM F2182-19e2 (2019)).

Heat Sink. A device designed to carry heat away from an object to which it is connected. Example: Ice pack (ACR Committee on MR Safety (2020)).

Hot Spot. Inhomogeneities in the electrical properties of tissue may result in associated inhomogeneous power deposition which may lead to focal areas of increased temperature in the patient (Schaefer (2014)).

Hyperechoes. An rf pulse train in a fast spin echo sequence which uses a series of small flip angles with variable phase with intermittent full angle 180-

degree pulses. These are useful for reducing SAR (McRobbie (2020)).

I

IAC. See Intersocietal Accreditation Commission.

ICD. Implanted Cardiac Defibrillator. Like a pacemaker but also can deliver a shock to reestablish heart rhythm. Distinguishable on chest X-ray by areas of thicker leads and larger "can" containing the generator with battery plus a capacitor for defibrillation. Both MR Unsafe and MR Conditional devices are in use (Indik et al. (2017)).

ICNIRP. See International Commission on Non-Ionizing Radiation Protection.

IEC. See International Electrotechnical Commission.

Impedance (Z). (*MRSE*) A measure of the opposition of flow to AC current in a circuit (Halliday et al. (2014)).

Implants. Medical implants are increasingly common. Implants are labeled as to their suitability for MRI as Safe, Conditional, or Unsafe. The conditions of testing are provided as a guide, but greater or lesser extremes of imaging conditions compared to the product labeling could potentially be safe. The decision to scan an implant for which product labeling information is not available or a decision to scan an implant "off-label" should be made based on a risk-benefit assessment. Final determination whether or not to scan any implant rests with the Level 2 attending MR radiologist, the MR medical director, or designated Level 2 personnel using predetermined protocols. Assessment of risks includes consideration of the likelihood of and effects of forces and currents which might occur during scanning (ACR Committee on MR Safety (2020)). A useful reference providing information concerning various implants is Shellock (2020). A device with dimensions of less than 2 cm in all directions is expected to heat less than 2 °C over 60 minutes of exposure at 1.5 T and 3 T (ASTM F2182-19e2 (2019)).

Example: A Conditional implant's labeling indicates it is safe to scan at less than 1.5 T, a maximum spatial gradient of 20 T$/m$, a slew rate of less than 200 T$/m/s$, and a whole body SAR of less than 2 W/kg for 15 min of scanning per sequence. Your scanner is 1.5 T, with a maximum spatial gradient of 15 T$/m$ and a slew rate of 200 T$/m/s$. Is it safe to scan the implant? Yes, the values of the parameters for your scanner in Normal Operating Mode are equal to or below the limits on the labeling. It may be possible to scan the device under more extreme conditions, but the scanner configuration is consistent with safe scanning per the device labeling.

Inadvertent Placement of Object in Magnet. In the event a potentially dangerous object is discovered in the patient during a scan, per the 2020 ACR Manual on MR Safety the course of action should be dictated by the clinical scenario and assessment of the likelihood for injury should the examination be allowed to proceed. Once the examination is terminated, either immediately or after additional imaging is obtained, the patient should be removed as slowly as possible along a path as much parallel to the bore as possible and instructed not to move until the patient is in a region of relatively low field < 5 G. The process of removal from the scanner described above is to minimize changes in force related to the position of the object with respect to the field and minimize changes in force related to velocity (Lorentz forces) (ACR Committee on MR Safety (2020)).

Inductance (L). (*MRSE*) Measured in Henries. The inductance L is defined as the proportionality related to how the emf changes as the current in a circuit changes with time (Halliday et al. (2014)).

$$emf = -L\frac{dI}{dt} \tag{I.1}$$

Inductance is measured in Henries (H).

Induction. See Magnetic Induction.

Inductor. (*MRSE*) Conducting coil which stores energy in a magnetic field when there is a current running through the coil (Halliday et al. (2014)).

Infection Control. Infection control in the MRI suite should proceed in much the same way as the rest of the facility. Cleaning equipment which is safe in the MRI environment must be used, and facility design should facilitate infection control (ACR Committee on MR Safety (2020)).

International Commission on Non-Ionizing Radiation Protection (ICNIRP). (*MRSE*) Independent international body providing scientific advice and guidance on the health and environmental effects of non-ionizing radiation including rf as well as static magnetic and electric fields.

Addresses standards pertaining to the static magnetic field, *emf* from motion in the static field 0-100 Hz, time varying gradients 0.5-5 kHz, applied rf 8.5-500 MHz, and laser radiation. Pertinent publications include:

2020 Guidelines for Limiting Exposure to Electromagnetic Fields (International Commission on Non-Ionizing Radiation Protection (2020))

2017 Statement on Diagnostic Devices using Non-ionizing Radiation: Existing Regulations and Potential Health Risks (International Commission on Non-Ionizing Radiation Protection (2017))

2014 Guidelines on Induced Electric Fields (International Commission on Non-Ionizing Radiation Protection (2014))

2013 Guidelines on Limits of Exposure to Laser Radiation (International Commission on Non-Ionizing Radiation Protection (2013))

2010 Guidelines on Limiting Exposure to Low Frequency Time Varying Electric and Magnetic Fields (International Commission on Non-Ionizing Radiation Protection (2010))

2009 Guidelines on Limits of Exposure to Static Magnetic Fields (International Commission on Non-Ionizing Radiation Protection (2009b))

2004 Statement on Protecting Patients During MRI Procedures (International Commission on Non-Ionizing Radiation Protection (2004), International Commission on Non-Ionizing Radiation Protection (2009a))

1998 Guidelines for Limiting Exposure to Low Frequency Time Varying Electric, Magnetic, and Electromagnetic Fields (International Commission on Non-Ionizing Radiation Protection (1998))

International Electrotechnical Commission (IEC). (*MRSE*) International nonprofit body which develops standards regarding electric and electronic devices and systems. The standard relevant to MRI safety is IEC 60601-2-33 published in 2010 with amendments in 2013 and 2015 (International Electrotechnical Commission (2010)). In addition, IEC 60825-1 2014 is relevant to laser safety (International Electrotechnical Commission (2014)) and IEC 61672-1 2013 is relevant to acoustic noise (International Electrotechnical Commission (2013)).

International Organization for Standardization (ISO). (*MRSE*) Independent international organization which develops voluntary consensus standards. The relevant standards regarding MRI Safety are ISO TS 10974 regarding active implanted devices (International Organization for Standardization (2018)) and ISO 226:2003 regarding acoustic noise (International Organization for Standardization (2013)).

Intersocietal Accreditation Commission (IAC) One of the accrediting organizations designated to accredit MRI sites under MIPPA (Clausen et al. (2018)).

Inversion Recovery (IR). A pulse sequence which utilizes an initial 180-degree inversion pulse followed by a 90-degree pulse (Brown et al. (2014)). Inversion recovery sequences have higher SAR than their standard counterparts by virtue of the extra 180-

degree inversion pulse. For example, a FSEIR will have a higher SAR than the corresponding FSE, all other parameters being equal.

ISO. See International Organization for Standardization.

Isocenter. The center of the bore of the magnet. This is where the patient accessible $\mathbf{B}_0$ field is the strongest and is the nominal field of the scanner, for example 1.5 T. This is where the field is most homogeneous. The time varying gradient strength is zero at the isocenter. The applied rf is maximal at isocenter. The spatial gradient is zero at the isocenter. The greatest magnitude of the field inside the magnet is greater than the field strength at isocenter in order to achieve the nominal field at isocenter.

Isotropic. Having the same values in all directions. For example, a sphere has the same length in all directions. See Anisotropic.

J

JCAHO. Joint Commission on Accreditation of Healthcare Organizations. See TJC.

Joule. A unit of energy in the SI system (Halliday et al. (2014)).

(*MRSE*) $1\ J = 1\ kg\text{-}m^2/s^2$. Of particular relevance to MRI, it is the energy of one ampere of current passing through one ohm of resistance for one second (Halliday et al. (2014)).

Joule Heating. Heating produced by the passage of current through a resistive conductor. Example: heating of the elements in a toaster. Also called Ohmic heating or resistive heating (Griffiths (2017)). This is an important mechanism of heating in MRI.

Joule Heating Law. (*MRSE*) Relationship of power deposition to current and resistance (Griffiths (2017)).

$$P = I^2 R \tag{J.1}$$

where P is the deposited power, I is the current, and R is resistance.

More generally in terms of the electric field $\mathbf{E}$, current density $\mathbf{J}$, conductivity σ_c, and resistivity ρ_c (Zangwill (2013))

$$P = \int_V \mathbf{E} \cdot \mathbf{J} dV = \int_V \rho_c \mathbf{J} \cdot \mathbf{J} dV = \int_V \rho_c \left| \mathbf{J} \right|^2 dV \tag{J.2}$$

where Ohm's Law $\mathbf{J} = \sigma_c \mathbf{E} = \mathbf{E}/\rho_c$ has been used.

Example: How does the power deposited change in a material when the current is doubled? Because power depends on the square of the current the power will quadruple.

K

Kelvin (K). Unit of temperature (Halliday et al. (2014)). 0 K is the lowest possible temperature. To convert from Kelvin to Celsius

$$°\mathrm{C} = \mathrm{K} + 273.15 \tag{K.1}$$

L

Larmor Equation. (*MRSE*) Fundamental MRI equation relating the magnetic flux density to the angular frequency Brown et al. (2014),

$$\omega = \gamma B \tag{L.1}$$

where ω is angular frequency, γ is the gyromagnetic ratio for the observed nucleus, and B is the field strength. This is true for the static B_0 and rf B_1 fields giving

$$\omega_0 = \gamma B_0 \tag{L.2}$$

and

$$\omega_1 = \gamma B_1 \tag{L.3}$$

Note that based on the direction of precession for a positive gyromagnetic ratio and the static field along $+z$, $\omega_0 = -\gamma B_0$ with clockwise precession; however, the convention of dropping the minus sign is often used (Brown et al. (2014)).

It is also frequently useful to use the frequency form of the equation

$$\nu_0 = \gamma B_0 \tag{L.4}$$

in which $\gamma = 42.6$ MHz/T and ν_0 is in MHz and B_0 is in Tesla.

Laser. A laser is used for scouting the patient prior to the scan and may cause damage to the eyes if it shines directly into the patient's eyes. Eyes should be shielded from the laser.

(*MRSE*) The relevant publications pertaining to laser safety are IEC 60825-1:2007 (International Electrotechnical Commission (2014)) and 2013 ICNIRP Guidelines on Exposure to Laser Radiation (International Commission on Non-Ionizing Radiation Protection (2013)). Lasers are Class II FDA and IEC devices. The laser typically has a label with the wavelength and power limit such as 650 nm and $< 0.35\, mW$.

Lead Burn. Burn resulting from a conducting lead. This may result from magnetic induction in the case of a looped lead or direct interaction of the electric field with the lead as in the antenna effect (Dempsey et al. (2001), Nyenhuis (2014), Tsai et al. (2015)).

Lead Impedence. (*MRSE*) A measure of the opposition of flow to AC current in an electrical lead. In terms of MRI safety this affects the degree to which a lead can heat up. In addition, in the imaging of CIEDs the impedance of the interrogated leads must be within a certain tolerance for safe scanning of conditional devices. Lead impedance should be evaluated pre and post scanning (Indik et al. (2017)).

Lenz Force. Magnetic component of the Lorentz force. See Lorentz Force.

Lenz's Law. (*MRSE*) Qualitative law indicating that when eddy currents are induced by a magnetic field as a consequence of Faraday's Law the resulting field from the induced current opposes the initial magnetic field (Zangwill (2013)).

Level 1 MR Personnel. Individuals with MR safety training defined by the MRMD sufficient to maintain their own safety and not be a threat to others as they work within Zone III. Level 1 personnel may move freely through all Zones, but may not admit Non-MR personnel or be responsible for Non-MR personnel in Zone III or Zone IV. Level I MR Personnel must regularly and routinely work within Zone III in order for them to maintain Level 1 status. Level designation must be renewed with training annually (ACR Committee on MR Safety (2020)).

Level 2 MR Personnel. Individuals who have been extensively trained in MR safety, with safety topics including but not limited to safety issues regarding: static fields, applied rf fields, time varying gradient fields, cryogens, contrast, etc. Only Level 2 personnel can admit Non-MR personnel to and supervise them within Zone III and Zone IV. Those entering Zone IV must be screened by at least one Level 2 personnel. Level designation must be renewed with training annually (ACR Committee on MR Safety (2020)).

Ligand. A molecule which binds another atom, ion, or molecule. Example: A gadolinium ion is bound by an organic ligand to form a contrast agent (ACR Committee on Drugs and Contrast Media (2020)).

Linearly Polarized Coil. A design of MRI coil in which transmission and reception take place along a single axis. The power deposition is twice that and the signal to noise $\sqrt{2}$ that of a quadrature coil (Brown et al. (2014)). See Coil Polarization.

Liquid Helium. Liquified helium which is used as a cryogen to cool the main MRI scanner coil to superconducting temperatures. The boiling point of liquid helium is 4.2 K (-452 °F). The expansion ratio is around 750 (Faulkner (2016)). See Expansion Ratio.

Liquid Nitrogen. Particularly in older magnets liquid nitrogen may surround the liquid helium in a superconducting magnet. The boiling point of liquid nitrogen is 77.4 K (-320 °F). The expansion ratio is around 700. See Expansion Ratio.

Local SAR. Specific Absorption Rate (SAR) averaged over 10 g of tissue over a 6 minute time period (Schaefer (2014)).

Local Transmit Receive Coil. A coil also referred to as a transceiver coil which can locally both transmit and receive rf signals. Example: Head transmit/receive coil. This should be distinguished from a receive only coil which always uses the built in body coil to transmit over much of the central volume of the bore (Brown et al. (2014), Ansorge and Graves (2016)). The safety significance is that the only the volume of anatomy inside the local coil receives rf, as opposed to a receive only coil which is excited by the body coil over the anatomy covered by the built in body coil within the bore.

Loop Burn. A burn resulting from a current generated in a closed curvilinear conductor via Faraday's Law. This could be a looped monitoring lead or a loop created by the inner thighs of the patient touching with the loop between the lower thighs up and around the groin and back down (Tsai et al. (2015), Mandel et al. (2017)).

Lorentz Force. A force with an electric field mediated component and a magnetic field mediated component (Halliday et al. (2014), Zangwill (2013), Griffiths (2017)). In MRI safety, the component of most relevance is the magnetic component. The magnetic component of the Lorentz force is referred to in much of the MR safety literature as the Lenz force. A conductor, which does not have to be magnetic, experiences a Lorentz force as it moves through a magnetic field. Alternatively, the current flowing through a wire in an electric circuit may give rise to a Lorentz force. Example 1: A nonmagnetic oxygen tank will resist coming toward the magnet. The faster the object is moved, the stronger the force. Example 2: A patient with an MR Conditional implant could experience a sensation from being moved rapidly into the magnet, but not when moved slowly into the magnet. The force could erroneously lead to the conclusion that the object has a significant degree of magnetism. Example 3: Currents flowing through the gradient coils cause the coils to move, creating the noise heard during the scan. Example 4: A quarter cannot be spun next to the bore of the magnet. The quarter is not magnetic but is a conductor. The spinning motion gets stopped by the Lorentz force.

(*MRSE*) The Lorentz force law states that a positive charge q moving in an electric field $\mathbf{E}$ at some velocity $\mathbf{v}$ with a velocity component perpendicular to a magnetic field $\mathbf{B}$ will experience a force perpendicular to both the direction of the magnetic field and the component of the velocity perpendicular to the field. For example a charge moving along the x-axis in a magnetic field oriented along the z-axis will feel a force along the y-axis (Zangwill (2013)).

Therefore the Lorentz force is

$$\mathbf{F} = q(\mathbf{E} + \mathbf{v} \times \mathbf{B}) \tag{L.5}$$

and for the magnitude

$$F = qE + qvB \sin\theta \tag{L.6}$$

For many applications in MRI safety, only the magnetic component of the Lorentz force (Lenz force) is important and

$$\mathbf{F} = q\mathbf{v} \times \mathbf{B} \tag{L.7}$$

For a wire carrying a current I along a length vector $\mathbf{l}$ this can be written as

$$\mathbf{F} = I\mathbf{l} \times \mathbf{B} \tag{L.8}$$

The magnitude F is

$$F = IlBsin\theta \tag{L.9}$$

where θ is the angle between $\mathbf{l}$ and $\mathbf{B}$.

The Lorentz force is what causes the gradient motion which produces acoustic noise during a scan.

If a moving conductor, which does not have to be magnetic, moves through an inhomogeneous field as would occur moving a patient or object into the magnet, eddy currents may be induced in the conductor through Faraday's Law. Since this current represents moving charge in a magnetic field, there will be a Lorentz force on the object.

This may be important as it could exert potentially dangerous forces on an object if moved rapidly through the field, or could give the erroneous impression that the object is ferromagnetic.

Example: Suppose as shown in Figure L.1 below that a conducting square ring with length l and a resistance R traveling at a velocity v along $+x$ moves perpendicular the direction of a constant magnetic field $\mathbf{B}$. The magnetic field $\mathbf{B}$ is directed out of the page along z in Figure L.1(b). Suppose that the field begins and ends abruptly along the x-axis. Suppose the right portion of the ring is within the field but the left is not.

In Figure L.1 the conducting square is moving to the right along the x-axis. The field is changing within the ring since part of the square is in the field and part is not. The changing field induces a flow of charge (a current) in the square in the direction dictated by Lenz's Law. By the Lorentz force law there is an associated force in the direction of the orange arrow which opposes the motion of the object into the field.

By Faraday's Law the induced electromotive force is

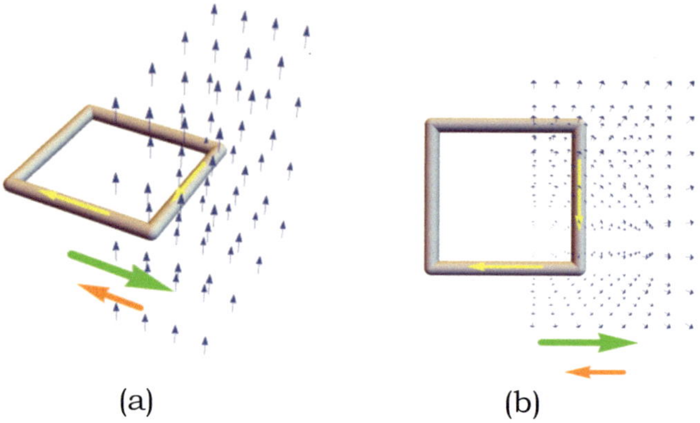

Figure L.1: The Lorentz Force on a Moving Square Conductor. (a) A conducting square moves in the direction x (green arrow) in a magnetic field (blue arrows) which generates a current (yellow arrows) with a resulting Lorentz force in the opposite direction of motion (orange arrow). Note that there is a portion of the square inside and outside the field. (b) Same as (a) viewed from above.

$$
\begin{aligned}
emf\, induced\, in\, ring &= \oint_C \mathbf{E} \cdot d\mathbf{l} = -\frac{\partial}{\partial t}\int_S \mathbf{B} \cdot d\mathbf{A} \\
&= -B\frac{\partial A}{\partial t} = -B\frac{\partial lx}{\partial t} \\
&= -Bl\frac{\partial x}{\partial t} = -Blv
\end{aligned}
\tag{L.10}
$$

Note that the magnitude of the current in the loop is given by Ohm's Law as

$$
I = \frac{V}{R} = \frac{Blv}{R} \tag{L.11}
$$

and flows in a clockwise direction by Lenz's Law.

The portions of the ring within the magnetic field will experience a Lorentz force from the current flowing in through the magnetic field. The portions of the current along the x and $-x$ direction are in opposite directions and will cancel each other out. The portions of the ring outside the field experience no Lorentz force. The force on the l portion of the ring in the field flowing in the $-y$ direction is

$$\begin{aligned}\mathbf{F} &= \frac{Blv}{R}\mathbf{l} \times \mathbf{B} = -\frac{Blv}{R} l\widehat{\mathbf{y}} \times B\widehat{\mathbf{z}} \\ &= -\frac{Blv}{R} lB \sin\theta\widehat{\mathbf{x}} = -\frac{B^2 l^2 v}{R}\widehat{\mathbf{x}}\end{aligned} \tag{L.12}$$

Therefore, there is a force directed along the $-x$ direction opposing the motion of the ring in the x direction. Note that the force increases with the square of B, the square of the dimension of the ring l, and the velocity v. This why a nonmagnetic conductor such as a nonferrous oxygen tank will experience a force if moved rapidly toward the scanner.

Now consider another case in which the square moves along $+x$ in an inhomogeneous field $B = B_z(x)$ with a constant field gradient of $\partial B/\partial x = \partial B_z(x)/\partial x$. The square ring in this case can be entirely within the field. It is assumed the field in this simple example can be described by

$$B = B(0) + x\frac{\partial B}{\partial x} \tag{L.13}$$

where $B(0)$ is the initial value far from the source of the field, so that

$$\frac{\partial B}{\partial t} = \frac{\partial x}{\partial t}\frac{\partial B}{\partial x} \tag{L.14}$$

The area of the square in the inhomogeneous field is constant and

$$\begin{aligned} emf\, induced\, in\, ring &= \oint_C \mathbf{E} \cdot d\mathbf{l} = -\frac{\partial}{\partial t}\int_S \mathbf{B} \cdot d\mathbf{A} \\ &= -A\frac{\partial B}{\partial t} = -A\frac{\partial x}{\partial t}\frac{\partial B}{\partial x} \\ &= -Av\frac{\partial B}{\partial x} \end{aligned} \tag{L.15}$$

The magnitude of the current in the loop is then again given by Ohm's Law as

$$I = \frac{V}{R} = \frac{Av}{R}\frac{\partial B}{\partial x} \tag{L.16}$$

and flows in a clockwise direction by Lenz's Law.

The ring will again experience a Lorentz force from the current flowing in the inhomogeneous magnetic field. The portions of the current along the x and $-x$ direction are in opposite directions and will cancel each other out. The force on the component of I flowing in the portion of the ring in the y and $-y$ directions are opposite, but will be less for the component at the smaller value of x. The force along $-x$ at the larger value of B is then

$$\begin{aligned}
\mathbf{F}_{-x} &= \frac{Av}{R}\frac{\partial B}{\partial x}\mathbf{l}\times\mathbf{B} = -\frac{Avl}{R}\frac{\partial B}{\partial x}\widehat{\mathbf{y}}\times B\widehat{\mathbf{z}} \\
&= -\frac{AvlB}{R}\frac{\partial B}{\partial x}\sin\theta\widehat{\mathbf{x}} = -\frac{Bvl^3}{R}\frac{\partial B}{\partial x}\widehat{\mathbf{x}} \\
&= -\widehat{\mathbf{x}}\frac{Bvl^3}{R}\frac{\partial B}{\partial x}
\end{aligned} \tag{L.17}$$

Using Eq L.13 and denoting the value of the force in the $-x$ direction as $\mathbf{F}_{-x}$

$$\begin{aligned}
\mathbf{F}_{-x} &= -\widehat{\mathbf{x}}\frac{Bvl^3}{R}\frac{\partial B}{\partial x} \\
&= -\widehat{\mathbf{x}}\frac{vl^3}{R}\frac{\partial B}{\partial x}\left(B(0) + x_2\frac{\partial B}{\partial x}\right)
\end{aligned} \tag{L.18}$$

with the larger value of x along the leading edge of the ring denoted x_2.

Similarly, there is a force along $+x$ at the trailing edge of the ring in the smaller portion of the field at x_1

$$\mathbf{F}_x = \widehat{\mathbf{x}}\frac{vl^3}{R}\frac{\partial B}{\partial x}\left(B(0) + x_1\frac{\partial B}{\partial x}\right) \tag{L.19}$$

Noting that the total force is

$$\mathbf{F} = \mathbf{F}_x + \mathbf{F}_{-x} \tag{L.20}$$

then

$$\mathbf{F} = -\widehat{\mathbf{x}}\frac{vl^3}{R}\left(\frac{\partial B}{\partial x}\right)^2\left(x_2 - x_1\right) \tag{L.21}$$

Since $x_2 > x_1$ and $l = x_2 - x_1$

$$\mathbf{F} = -\widehat{\mathbf{x}}\frac{vl^4}{R}\left(\frac{\partial B}{\partial x}\right)^2 \tag{L.22}$$

Therefore, in the case of a linear static field gradient there is a force directed along the $-x$ direction opposing the motion of the ring in the x direction. Note that the force increases with the square of $\partial B/\partial x$, the fourth power of the dimension of the ring l, and the velocity v.

M

Magnetic Braking. Use of the Lorentz force to stop an object. For example, a metallic conductor moving in a given direction toward the magnet as the field is increasing with distance will have eddy currents induced in it which in turn produce a Lorentz force opposing the motion of the object (Knoepfel (2000)).

Magnetic Dipole. A pair of magnetic poles (magnetic charges) separated by some distance (Zangwill (2013)).

Magnetic Dipole Field. A field from magnetic north and south poles oriented along the z-axis separated by a distance (Zangwill (2013)).

$$\mathbf{B}_d(r,\theta) = \frac{\mu_0 m}{4\pi r^3}(2\cos\theta\widehat{\mathbf{r}} + \sin\theta\widehat{\boldsymbol{\theta}}) \qquad \text{(M.1)}$$

where μ_0 is the magnetic permeability of free space, m is the magnitude of the magnetic moment, θ is the angle between the z-axis and the observation point, and $\widehat{\mathbf{r}}$ and $\widehat{\boldsymbol{\theta}}$ are the unit vectors in spherical coordinates.

Magnetic Displacement Force (F_m). Translational force which tries to pull a magnetic object toward the scanner (Schenck (2000), Schenck and Kangarlu (2014)). This force is responsible for the missile effect. The force on an unsaturated object is proportional to the force product of the field strength and the change in field strength with the change in distance (spatial gradient), that is

$$F_m \propto B_0 \frac{dB_0}{dz} \quad \text{(M.2)}$$

Once an object becomes saturated then

$$F_m \propto \frac{dB_0}{dz} \quad \text{(M.3)}$$

Consequently, the force is only active in an inhomogeneous field. Near the center of the bore the field is homogeneous, and the translational force is close to zero, as opposed to the magnetic torque, which is maximal at the center of the bore. Note the displacement force is also zero far from the magnet where the field far from the magnet is zero. The spatial gradient and the displacement force are usually maximal near the entrance to the bore of the scanner

(*MRSE*) More specifically, for the simple one dimensional case, as would be a good approximation along the axis of the bore, for an object with magnetic susceptibility much less than 1, an unsaturated object would have a displacement force

$$F_m = \frac{\chi_m V}{\mu_0} B_0 \frac{dB_0}{dz} \quad \text{(M.4)}$$

where F_m is the force, χ_m is the magnetic susceptibilty of the object (assuming scalar magnetic susceptibility), μ_0 is the magnetic permeability of free space, V is the volume of the object, and B_0 is the field strength. The derivative $dB_0(z)/dz = dB_0/dz$ describing the change in the field with distance is the spatial gradient.

Note that the force increases with the volume of the object, the magnetic susceptibility, the field strength, and the spatial gradient. The maximum displacement force on an unsaturated object occurs when the product $B_0 dB_0/dz$ is maximized (the maximum force product), typically near where the value of dB_0/dz is maximized (the maximum spatial gradient). For a magnetically saturated object, the field dependence of the force only depends upon dB_0/dz (ASTM F2052-15 (2015)).

For the even more general case the force can be calculated from the magnetic potential energy. See Magnetic Potential Energy.

For a saturated magnetic object with magnetic moment $\mathbf{m}_S$, the magnetic moment is not a function of the magnetic flux density $\mathbf{B}_0$, and the potential energy U is given by (Schenck (2000), Schenck and Kangarlu (2014))

$$U = -\mathbf{m}_S \cdot \mathbf{B}_0 \tag{M.5}$$

To compute the magnetic displacement force for a saturated object

$$\mathbf{F}_m = -\nabla U = \nabla \left(\mathbf{m}_S \cdot \mathbf{B}_0 \right) \tag{M.6}$$

Using the vector identity

$$\begin{aligned} \nabla (\mathbf{a} \cdot \mathbf{b}) = {} & (\mathbf{a} \cdot \nabla) \mathbf{b} + (\mathbf{b} \cdot \nabla) \mathbf{a} \\ & + \mathbf{a} \times (\nabla \times \mathbf{b}) + \mathbf{b} \times (\nabla \times \mathbf{a}) \end{aligned} \tag{M.7}$$

where $\mathbf{a}$ and $\mathbf{b}$ are arbitrary vectors then

$$\nabla (\mathbf{m}_S \cdot \mathbf{B}_0) = (\mathbf{m}_S \cdot \nabla)\,\mathbf{B}_0 + (\mathbf{B}_0 \cdot \nabla)\,\mathbf{m}_S + \mathbf{m}_S \times (\nabla \times \mathbf{B}_0) + \mathbf{B}_0 \times (\nabla \times \mathbf{m}_S) \tag{M.8}$$

Noting that $\mathbf{m}_S$ does not change with position and that within the object $\nabla \times \mathbf{B}_0 = 0$ then (Boyer (1988))

$$\mathbf{F}_m = -\nabla U = \nabla (\mathbf{m}_S \cdot \mathbf{B}_0) = (\mathbf{m}_S \cdot \nabla)\,\mathbf{B}_0 \tag{M.9}$$

The case for an induced magnetic moment $\mathbf{m}_U$ in an unsaturated object in an external magnetic field $\mathbf{B}_0$ is more complex since the magnetization now depends on the external field (Schenck (2000), Schenck and Kangarlu (2014)). Starting with the magnetic potential energy U for an unsaturated object (See Magnetic Potential Energy)

$$U = -\frac{\mathbf{m}_U \cdot \mathbf{B}_0}{2} \tag{M.10}$$

Using the definition of magnetization $\mathbf{M}$ and its relationship to the magnetic flux density $\mathbf{B}_0$ (See Magnetization)

$$\mathbf{m}_U = \mathbf{M}V = \boldsymbol{\chi}_e \mathbf{H}_0 V = \frac{\boldsymbol{\chi}_e V \mathbf{B}_0}{\mu_0} \tag{M.11}$$

where V is the volume of the object, $\mathbf{H}_0$ is the applied static magnetic field, $\boldsymbol{\chi}_e$ is the magnetic susceptibility tensor in the scanner frame, and μ_0 is the magnetic permeability of free space.

The magnetic potential energy can then be written

$$U = -\frac{\chi_e V \mathbf{B}_0 \cdot \mathbf{B}_0}{2\mu_0} \tag{M.12}$$

To compute the force for an unsaturated object

$$\mathbf{F}_m = -\nabla U$$

$$= \frac{\chi_e V}{2\mu_0}\nabla(\mathbf{B}_0 \cdot \mathbf{B}_0) = \frac{\chi_e V}{2\mu_0}\nabla(B_0^2) = \frac{\chi_e V}{\mu_0} B_0 \nabla B_0 \tag{M.13}$$

In addition, the vector identity from Eq. M.7 is in this case

$$\nabla(\mathbf{B}_0 \cdot \mathbf{B}_0) = (\mathbf{B}_0 \cdot \nabla)\mathbf{B}_0 + (\mathbf{B}_0 \cdot \nabla)\mathbf{B}_0 + \mathbf{B}_0 \times (\nabla \times \mathbf{B}_0) + \mathbf{B}_0 \times (\nabla \times \mathbf{B}_0) \tag{M.14}$$

Again $\nabla \times \mathbf{B}_0 = 0$, but this time $\mathbf{m}_U$ is a function of position since it is a function of $\mathbf{B}_0$ so that

$$\nabla(\mathbf{B}_0 \cdot \mathbf{B}_0) = 2(\mathbf{B}_0 \cdot \nabla)\mathbf{B}_0 \tag{M.15}$$

Therefore the force may also be written as (ASTM F2052-15 (2015))

$$\mathbf{F}_m = \frac{\chi_e V}{2\mu_0}\nabla(\mathbf{B}_0 \cdot \mathbf{B}_0) = \frac{\chi_e V}{\mu_0}(\mathbf{B}_0 \cdot \nabla)\mathbf{B}_0 \tag{M.16}$$

Following the example given by (Schenck (2000) and Schenck and Kangarlu (2014)) the z component of $\mathbf{F}_m$ (F_{m_z}) is calculated for an ellipsoid with magnetization induced by the static field when the long axis of the ellipsoid, which is the x-axis in the principle axis system of the ellipsoid, is rotated around the y-axis by θ degrees from the scanner (lab) frame ($\beta = \theta - \pi/2$). It is further assumed that $\mathbf{B}_0 = \big(0, 0, B_0(z)\big) = \big(0, 0, B_0\big)$ and magnetic susceptibility $\chi_{xx} = \chi_{yy} = \chi_{zz} = \chi_m$ and demagnetizing factors $N_{xx} = N_c$ and $N_{yy} = N_{zz} = N_a = N_b$.

Since the principle axis system of the magnetic susceptibility tensor is not collinear with that of the magnetic field, an appropriate rotation of the magnetic susceptibility tensor in the principle axis system $\boldsymbol{\chi}_p$ must be performed using the relevant Euler rotation matrix $\mathbf{R}$ (See Magnetic Susceptibility Tensor)

$$\mathbf{F}_m = \frac{\boldsymbol{\chi}_e V}{\mu_0} B_0 \nabla B_0 = \frac{\mathbf{R}\boldsymbol{\chi}_p \mathbf{R}^{-1} V}{\mu_0} B_0 \nabla B_0 \quad \text{(M.17)}$$

After calculation (Schenck (2000), Schenck and Kangarlu (2014)) the z-component of the displacement force F_{m_z} is

$$F_{m_z} = \frac{\chi_m V}{2\mu_0} B_0 \frac{\partial B_0}{\partial z} \left[\frac{\cos^2\theta}{1 + \chi_m N_c} + \frac{\sin^2\theta}{1 + \chi_m N_a} \right] \quad \text{(M.18)}$$

For $\theta = 0$ or $\theta = \pi/2$

$$\chi_e = \frac{\chi_m}{1 + N_{ii}\chi_m} \tag{M.19}$$

where $N_{ii} = N_c$ or $N_{ii} = N_a$ for $\theta = 0$ or $\theta = \pi/2$, respectively.

Note that for the condition $\mathbf{B}_0 = \big(0, 0, B_0(z)\big) = \big(0, 0, B_0\big)$ and for $\theta = 0$ or $\theta = \pi/2$

$$F_m = \frac{\chi_e V}{\mu_0} B_0 \frac{dB_0}{dz} \tag{M.20}$$

These equations then allow the several useful relations to be derived (ASTM F2052-15 (2015))

$$\frac{F_{m2}}{F_{m_1}} = \frac{B_{0_2}\left|\nabla B_0\right|_2}{B_{0_1}\left|\nabla B_0\right|_1} \tag{M.21}$$

and employing the relationship

$$F_m = F_g \tan\alpha \tag{M.22}$$

it can be shown (ASTM F2052-15 (2015))

$$\left|\nabla B_0\right|_2 = \left|\nabla B_0\right|_1 \frac{B_{0_1}\tan\alpha_2}{B_{0_2}\tan\alpha_1} \tag{M.23}$$

In the case of a magnetically saturated object, the magnetic moment $\mathbf{m}$ does not vary with $\mathbf{B}_0$. In this case for $\mathbf{B}_0 = \big(0, 0, B_0(z)\big) = \big(0, 0, B_0\big)$ and saturated magnetic moment $\mathbf{m}_S$

$$\mathbf{F}_m = \left(\mathbf{m}_S \cdot \nabla\right)\mathbf{B}_0 = m_{S_z}\frac{dB_0}{dz} = m_S\frac{dB_0}{dz} \tag{M.24}$$

where m_{S_z} is the z-component of the saturated magnetic moment. For the saturated case the force ratio at two different spatial gradients becomes (ASTM F2052-15 (2015))

$$\frac{F_{m_2}}{F_{m_1}} = \frac{|\nabla B_0|_2}{|\nabla B_0|_1} \tag{M.25}$$

It is also instructive to consider how the magnetic displacement force varies with distance from the source of the field. If a single current carrying loop of radius R_c carrying current I centered at z' or Helmholz coil configuration at $z' = \pm R_c/2$ is considered (See Biot-Savart Law), the field falls off along the z-axis as $1/z^3$

$$\frac{\partial B(z)}{\partial z} = \frac{3\mu_0 I R_c^2 (z - z')}{2\left[R_c^2 + (z - z')^2\right]^{5/2}} \tag{M.26}$$

If the distance is significantly greater than the radius of the loop

$$\frac{\partial B(z)}{\partial z} = \frac{3\mu_0 I R_c^2}{2(z - z')^4} \tag{M.27}$$

Likewise for the Helmholtz pair

$$\begin{aligned}\frac{\partial B_h(z)}{\partial z} &= 24\mu_0 I R_c^2 \left[\frac{R_c - 2z}{(5R_c^2 - 4R_c z + 4z^2)^{5/2}}\right] \\ &+ 24\mu_0 I R_c^2 \left[\frac{R_c + 2z}{(5R_c^2 + 4R_c z + 4z^2)^{5/2}}\right]\end{aligned} \tag{M.28}$$

and again assuming that the distance is much greater than the radius of the loops

$$\frac{\partial B_h(z)}{\partial z} = \frac{-3I\mu_0}{z^4} \tag{M.29}$$

For a saturated object

$$F_m \propto \frac{\partial B_z}{\partial z} \propto \frac{1}{z^4} \tag{M.30}$$

while for an unsaturated object

$$F_m \propto B_z \frac{\partial B_z}{\partial z} \propto \frac{1}{z^7} \tag{M.31}$$

The relevant parameters for a set of Helmholtz coils are plotted in Figures M.1 - M.3.

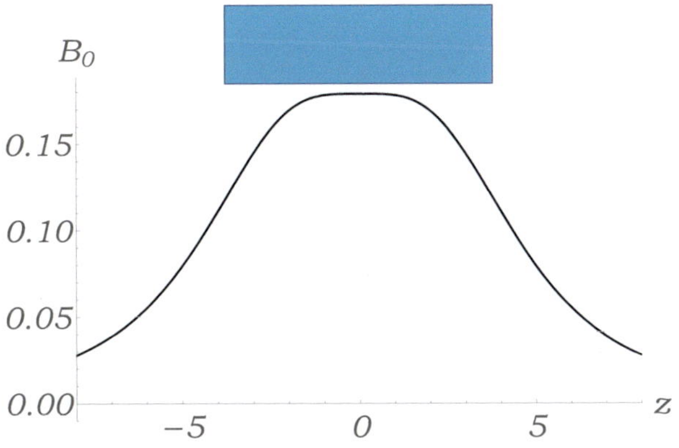

Figure M.1: Static field for a Helmholtz coil pair. The blue rectangle indicates the portion of the field as would be analogously present in the bore of an MRI scanner.

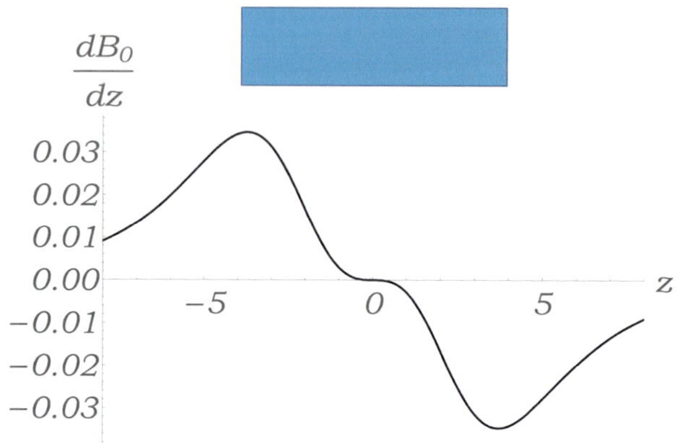

Figure M.2: Spatial gradient for a Helmholtz coil pair. The blue rectangle indicates the portion of the field as would be analogously present in the bore of an MRI scanner.

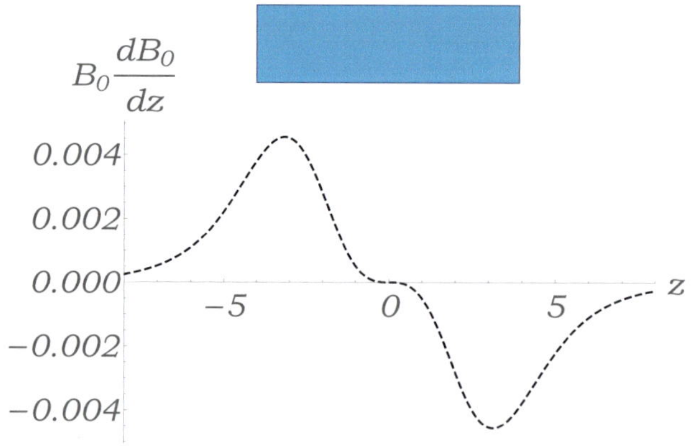

Figure M.3: Force product for a Helmholtz coil pair. The blue rectangle indicates the portion of the field as would be analogously present in the bore of an MRI scanner.

Note that in the center of the coils in Figure M.1 the field is homogeneous and dB_0/dz is zero, but at the edges of the coil configuration the field abruptly drops off and dB_0/dz changes rapidly and nonlinearly. This same information is also depicted as a plot of dB_0/dz with distance in Figure M.2. In Figure M.3 the force product is given as a function of distance. Note that this has an even sharper zone of transition.

The situation depicted in Figures M.1 - M.3 is for no shielding for a distance along the z-axis $r = z$. In the presence of active shielding, the field may fall off on the order of $1/r^5$. In principle then, for a saturated object, the force falls off as $1/r^6$ and for an unsaturated object $1/r^{11}$! In reality, for an actual scanner, there is not a simple analytical expression. However, force products and associated forces falling off on the order of $1/r^{10}$ may be present near the face of the bore in the region of the maximum spatial gradient. It is therefore easy to see how an object could not experience much force at one point in the room then suddenly be uncontrollable only a short distance closer to the magnet.

Magnetic Field (H). Region around objects where a magnetic force acts. Magnetic fields may be produced by electric currents or by magnetic materials. The magnetism of magnetic materials may be modeled by considering electron motion, however (Coey (2010)). See **H**.

Magnetic Flux Density (B). (*MRSE*) The total magnetic field including any external field and separate matter in the field (Zangwill (2013)). See **B**.

Magnetic Induction. (*MRSE*) Another name for Magnetic Flux Density. Also refers to the process of inducing currents through Faraday's Law (Zangwill (2013)). See **B**.

Magnetic Moment (m). An elementary average atomic magnetization equivalent to a small magnetic dipole.

(*MRSE*) Due to electron orbital and spin angular momentum or nuclear spin angular momentum, each atom or nucleus can be thought of as being a current loop which produces a magnetic field and thus acts like a tiny bar magnet which has an associated magnetic moment (Zangwill (2013)).

The magnetic moment $\mathbf{m}$ for a volume V is related to the volume magnetization $\mathbf{M}$ by

$$\mathbf{M} = \frac{d\mathbf{m}}{dV} \tag{M.32}$$

Consequently for a magnetically homogeneous sample

$$\mathbf{m} = \int_V \mathbf{M} dV = \mathbf{M}V \tag{M.33}$$

Magnetic Permeability (μ). (*MRSE*) Constant of proportionality linking the concept of the Magnetic Flux Density with the concepts of Magnetic Field and Magnetization. Magnetic Permeability is related to the Magnetic Permeability of Free Space and the Relative Magnetic Permeability (Zangwill (2013)). See **B**.

Magnetic Permeability of Free Space (μ_0). (*MRSE*) A constant relating $\mathbf{B}_0$ and $\mathbf{H}_0$. $\mu_0 = 1.257 \times 10^{-6}\ H/m$ (Zangwill (2013)).

Magnetic Potential Energy (U). (*MRSE*) The magnetic potential energy differs depending upon whether the object is saturated or unsaturated.

The magnetic potential energy U between $\mathbf{B}$ and the magnetic moment $\mathbf{m}$ is considered. It can be shown that (Barrett and Macdonald (1999), Wasserman (2011), Zangwill (2013), Sackett (2018))

$$dU = -\mathbf{m} \cdot d\mathbf{B} \tag{M.34}$$

so that in bringing a magnetic object into an external field at a flux density of $\mathbf{B}_0$ (Zangwill (2013))

$$U = -\int_0^{\mathbf{B}_0} \mathbf{m} \cdot d\mathbf{B} \tag{M.35}$$

For a saturated magnetic object with magnetic moment $\mathbf{m}_S$, the magnetic moment is not a function of $\mathbf{B}_0$ and (Schenck (2000), Schenck and Kangarlu (2014))

$$U = -\int_0^{\mathbf{B}_0} \mathbf{m}_S \cdot d\mathbf{B} = -\mathbf{m}_S \cdot \mathbf{B}_0 \tag{M.36}$$

The case for an induced magnetization in an unsaturated object with magnetic moment $\mathbf{m}_{U'}$ in an external magnetic field is more complex since the magnetization now depends on the external field

$$U = -\int_0^{\mathbf{B}_0} \mathbf{m}_{U'} \cdot d\mathbf{B} = -\int_0^{\mathbf{B}_0} \frac{\boldsymbol{\chi}_e V \mathbf{B} \cdot d\mathbf{B}}{\mu_0}$$
$$= -\frac{\boldsymbol{\chi}_e V \mathbf{B}_0 \cdot \mathbf{B}_0}{2\mu_0} \tag{M.37}$$

where the definitions of magnetization $\mathbf{M}$ and its relationship to the magnetic flux density $\mathbf{B}_0$

$$\int_V \mathbf{M} dV = \mathbf{M}V = \mathbf{m} \tag{M.38}$$

and

$$\mathbf{M} = \boldsymbol{\chi}_e \mathbf{H}_0 = \frac{\boldsymbol{\chi}_e \mathbf{B}_0}{\mu_0} \tag{M.39}$$

have been used. See Magnetization. $\boldsymbol{\chi}_e$ is the magnetic susceptibility tensor in the laboratory (scanner) frame, V is the volume of the object, $\mathbf{H}_0$ is the applied static magnetic field, $\mathbf{M}$ is the magnetization, and μ_0 is the magnetic permeability of free space. The magnetic potential energy for an unsaturated object can also be written as (Schenck (2000), Schenck and Kangarlu (2014))

$$U = -\frac{\mathbf{m}_U \cdot \mathbf{B}_0}{2} \tag{M.40}$$

for the induced magnetic moment $\mathbf{m}_U$ arising from a field going from a magnitude of 0 to B_0.

Following the example given by Schenck (2000) and Schenck and Kangarlu (2014)) the potential energy

U is calculated for an ellipsoid with magnetization induced by the static field when the long axis of the ellipsoid, which is the x-axis in the principle axis system of the ellipsoid, is rotated around the y-axis by θ degrees from the scanner (lab) frame ($\beta = \theta - \pi/2$). It is further assumed that $\mathbf{B}_0 = \big(0, 0, B_0(z)\big) = \big(0, 0, B_0\big)$ and magnetic susceptibility $\chi_{xx} = \chi_{yy} = \chi_{zz} = \chi_m$ and demagnetizing factors $N_{xx} = N_c$ and $N_{yy} = N_{zz} = N_a = N_b$.

Since the principle axis system of the magnetic susceptibility tensor is not collinear with that of the magnetic field, an appropriate rotation of the magnetic susceptibility tensor in the principle axis system $\boldsymbol{\chi}_p$ must be performed using the relevant Euler rotation matrix $\mathbf{R}$ (See Magnetic Susceptibility Tensor)

$$U = -\frac{\boldsymbol{\chi}_e V \mathbf{B}_0 \cdot \mathbf{B}_0}{2\mu_0} = -\frac{\mathbf{R}\boldsymbol{\chi}_p \mathbf{R}^{-1} V \mathbf{B}_0 \cdot \mathbf{B}_0}{2\mu_0}$$

$$= -\frac{\chi_m V B_0^2}{2\mu_0}\left[\frac{\cos^2\theta}{1+\chi_m N_c} + \frac{\sin^2\theta}{1+\chi_m N_a}\right] \tag{M.41}$$

Magnetic Saturation. (*MRSE*) When the magnetic dipoles in an object become aligned to such an extent that no more alignment and hence magnetization results from increasing the applied external field, the object is said to be saturated (Fernow (2016)). Saturation magnetizations have been determined for different elements and compounds, for example the saturtation magnetizations of iron, nickel, and cobalt are 2.15 T, 0.61 T, and 1.81 T, respectively (Coey (2010)).

Magnetic Scalar Potential (Ψ). (*MRSE*) A magnetic scalar potential can be used to calculate the magnetic field **B** via (Zangwill (2013))

$$\mathbf{B} = -\nabla\Psi \tag{M.42}$$

This approach is applicable when there is a static or quasistatic problem in which a current density is confined to a surface dividing current free volumes (Zangwill (2013)).

Magnetic Susceptibility (χ). A property of a substance which is a measure of its response to an external magnetic field and how well it becomes magnetized in an external magnetic field. The degree of magnetization determines in part the extent to which the object will interact with the magnetic field of the scanner. Ferromagnetic materials have high positive values of magnetic susceptibility, paramagnetic substances have small positive values of magnetic susceptibility, and diamagnetic substances have small negative values of magnetic susceptibility (Jiles (2015)). Ferromagnetic materials pose significant risk from associated forces from the external magnetic field.

(*MRSE*) The magnetic susceptibility is in general a tensor quantity $\boldsymbol{\chi}_m$ which assuming there is no significant demagnetizing field is in its principle axis system given by

$$\boldsymbol{\chi}_m = \begin{pmatrix} \chi_{xx} & 0 & 0 \\ 0 & \chi_{yy} & 0 \\ 0 & 0 & \chi_{zz} \end{pmatrix} \tag{M.43}$$

For many materials and situations the tensor may be described by a scalar (Zangwill (2013)). That is to say $\chi_{xx} = \chi_{yy} = \chi_{zz} = \chi_m$ and

$$\boldsymbol{\chi}_m = \chi_m \begin{pmatrix} 1 & 0 & 0 \\ 0 & 1 & 0 \\ 0 & 0 & 1 \end{pmatrix} \tag{M.44}$$

Consequently

$$\mathbf{M} = \boldsymbol{\chi}_m \mathbf{H} = \chi_m \mathbf{H} \tag{M.45}$$

and the magnetic susceptibility is described by a single scalar χ_m.

Magnetic Susceptibility Tensor ($\boldsymbol{\chi}$). (*MRSE*) Magnetic susceptibility may often be described by a scalar. In general, however, the susceptibility is different in different directions in a material and hence described as a tensor (Zélis et al. (2017)). Two magnetic susceptibility tensors can be defined in terms of the magnetization **M**

$$\mathbf{M} = \boldsymbol{\chi}_p \mathbf{H}_0 \tag{M.46}$$

and

$$\mathbf{M} = \boldsymbol{\chi}_m \mathbf{H}_{in} \tag{M.47}$$

$\boldsymbol{\chi}_m$ is defined with respect to the field inside the object $\mathbf{H}_{in}$ while $\boldsymbol{\chi}_p$ is defined with respect to the applied field $\mathbf{H}_0$.

The total field $\mathbf{H}_{in}$ inside a body in a magnetic field is

$$\mathbf{H}_{in} = \mathbf{H}_0 + \mathbf{H}_d = \mathbf{H}_0 - \mathbf{N} \cdot \mathbf{M} \tag{M.48}$$

where $\mathbf{N}$ is the demagnetizing tensor (See Demagnetizing Tensor). Multiplying both sides by $\boldsymbol{\chi}_m$ and $\boldsymbol{\chi}_p$ gives

$$\boldsymbol{\chi}_p\boldsymbol{\chi}_m\mathbf{H}_{in} = \boldsymbol{\chi}_p\boldsymbol{\chi}_m\mathbf{H}_0 - \boldsymbol{\chi}_p\boldsymbol{\chi}_m\mathbf{N} \cdot \mathbf{M} \tag{M.49}$$

Noting that $\boldsymbol{\chi}_m$, $\boldsymbol{\chi}_p$, and $\mathbf{N}$ commute in their principle axis system

$$\boldsymbol{\chi}_p\mathbf{M} = \boldsymbol{\chi}_m\mathbf{M} - \boldsymbol{\chi}_p\boldsymbol{\chi}_m\mathbf{N} \cdot \mathbf{M} \tag{M.50}$$

then solving for the components of $\boldsymbol{\chi}_p$ in the principle axis system in the presence of a demagnetizing field, the diagonal elements of the susceptibility tensor take the form (Zangwill (2013), Zélis et al. (2017))

$$\chi_{p_{ii}} = \frac{\chi_{ii}}{1 + \chi_{ii}N_{ii}} \tag{M.51}$$

where χ_{ii}, $\chi_{p_{ii}}$ and N_{ii} are the diagonal elements of $\boldsymbol{\chi}_m$, $\boldsymbol{\chi}_p$, and $\mathbf{N}$, respectively. Therefore, in the principle axis system of the susceptibility interaction

$$\boldsymbol{\chi}_p = \begin{pmatrix} \dfrac{\chi_{xx}}{1+\chi_{xx}N_{xx}} & 0 & 0 \\ 0 & \dfrac{\chi_{yy}}{1+\chi_{yy}N_{yy}} & 0 \\ 0 & 0 & \dfrac{\chi_{zz}}{1+\chi_{zz}N_{zz}} \end{pmatrix} \tag{M.52}$$

Note that if the magnetic field is applied along a principle axis of the object, then the magnetization $\mathbf{M}$ in a field $\mathbf{H}_0$ is

$$\mathbf{M} = \boldsymbol{\chi}_p \mathbf{H}_0 \tag{M.53}$$

and

$$M = M_i = \chi_{e_i} H_i = \frac{H_i \chi_{ii}}{1+\chi_{ii}N_{ii}} \tag{M.54}$$

where χ_{ii} are the elements of the susceptibility tensor in its principle axis system in the absence of a significant demagnetizing field. If, however, the applied magnetic field is not along a principle axis, then the tensor must be rotated from the principle axis system into the frame of the applied magnetic field (laboratory frame) using the transformation

$$\boldsymbol{\chi}_e = \mathbf{R}\boldsymbol{\chi}_p\mathbf{R}^{-1} \tag{M.55}$$

and then

$$\mathbf{M} = \boldsymbol{\chi}_e \mathbf{H}_0 \tag{M.56}$$

where $\boldsymbol{\chi}_e$ is the magnetic susceptibility tensor in the scanner (lab) frame, $\boldsymbol{\chi}_p$ is the magnetic susceptibility tensor in the principle axis frame, and $\mathbf{R}$ is the Euler rotation matrix. See Euler Rotation Matrix.

Magnetic Torque ($\boldsymbol{\tau}_m$). A force which tries to rotate an object in the magnetic field (Schenck (2000), Schenck and Kangarlu (2014)). This is strongest where the magnetic field is the strongest, which for the patient accessible region is at the center of the bore. For an unsaturated object the torque τ_m in magnetic field B_0 is

$$\tau_m \propto B_0^2 \tag{M.57}$$

For a saturated object

$$\tau_m \propto B_0 \tag{M.58}$$

(*MRSE*) The torque on an object with magnetic moment $\mathbf{m}$ in a field $\mathbf{B}$ at position $\mathbf{r}$ is given by (Zangwill (2013))

$$\boldsymbol{\tau}_m(\mathbf{r}) = \mathbf{m}(\mathbf{r}) \times \mathbf{B}(\mathbf{r}) + \mathbf{r} \times \mathbf{F}_m(\mathbf{r}) \tag{M.59}$$

where $\mathbf{F}_m$ is the magnetic displacement force (See Magnetic Displacement Force). For the simplifying assumption of a uniform magnetic field such as at the center of the bore

$$\boldsymbol{\tau}_m = \mathbf{m} \times \mathbf{B}_0 \tag{M.60}$$

which implies that the magnitude of the torque for an angle θ between the magnetic moment and flux density vectors is

$$\tau_m = mB_0 \sin\theta \tag{M.61}$$

For a saturated object the magnetic moment $\mathbf{m}_S$ is independent of the applied field and

$$\boldsymbol{\tau}_m = \mathbf{m}_S \times \mathbf{B}_0 \tag{M.62}$$

For an unsaturated object the magnetic moment $\mathbf{m}_U$ is a function of the applied field. From the definition of $\mathbf{m}$, together with the effects of susceptibility and demagnetization on an unsaturated object (See Magnetization)

$$\begin{aligned}\boldsymbol{\tau}_m &= \mathbf{m}_U \times \mathbf{B}_0 = \mathbf{M}V \times \mathbf{B}_0 = \boldsymbol{\chi}_e \mathbf{H}_0 V \times \mathbf{B}_0 \\ &= \boldsymbol{\chi}_e \frac{V\mathbf{B}_0}{\mu_0} \times \mathbf{B}_0 = \mathbf{R}\boldsymbol{\chi}_p \mathbf{R}^{-1} \frac{V\mathbf{B}_0}{\mu_0} \times \mathbf{B}_0\end{aligned} \tag{M.63}$$

where V is the volume of the object, $\mathbf{H}_0$ is the applied static magnetic field, $\boldsymbol{\chi}_e$ is the magnetic susceptibility tensor in the scanner frame, and μ_0 is the magnetic permeability of free space. Since the principle axis system of the magnetic susceptibility tensor is not collinear with that of the magnetic field, an appropriate rotation of the magnetic susceptibility tensor in the principle axis system $\boldsymbol{\chi}_p$ must be performed using the relevant Euler rotation matrix $\mathbf{R}$. See Magnetic Susceptibility Tensor.

Following the example given by (Schenck (2000) and Schenck and Kangarlu (2014)) the above equation can then be used to calculate the torque, which will be in the y-direction, giving the y-component τ_{m_y}, assuming an ellipsoid with magnetization induced by the static field when the long axis of the ellipsoid, which is the x-axis in the principle axis system of the ellipsoid, is rotated around the y-axis by θ degrees from the scanner (lab) frame ($\beta = \theta - \pi/2$). It is further assumed that $\mathbf{B}_0 = \big(0, 0, B_0(z)\big) = \big(0, 0, B_0\big)$ and magnetic susceptibility $\chi_{xx} = \chi_{yy} = \chi_{zz} = \chi_m$ and demagnetizing factors $N_{xx} = N_c$ and $N_{yy} = N_{zz} = N_a = N_b$.

As previously given (Schenck (2000), Schenck and Kangarlu (2014))

$$\tau_{m_y} = \frac{\chi_m^2 V B_0^2}{\mu_0} \left[\frac{(N_c - N_a)}{(1 + \chi_m N_c)(1 + \chi_m N_a)} \right] \cos\theta \sin\theta \tag{M.64}$$

The maximum torque occurs where B_0 is maximal, which for the patient accessible regions of the scanner is at isocenter, and at an angle of $\theta = \pi/4$ (Schenck (2000), Schenck and Kangarlu (2014)).

In terms of the magnetic potential energy (Schenck (2000), Schenck and Kangarlu (2014))

$$\boldsymbol{\tau}_{m_y} = \frac{\partial U}{\partial \theta} \widehat{\mathbf{y}} \tag{M.65}$$

where U is the magnetic potential energy (See Magnetic Potential Energy).

Note that for a sphere $N_c - N_a = 0$ and $\tau_{m_y} = 0$. In addition if the object is aligned with the field $\tau_{m_y} = 0$.

It is also of interest to comment on how the torque changes with distance. Considering a current loop without shielding, the torque for a saturated object would vary as $1/r^3$, and for an unsaturated object $1/r^6$. For an actively shielded magnet with a field varying as $1/r^5$ to a first approximation a saturated object's torque would fall off as $1/r^5$ while an unsaturated object would have a steeper $1/r^{10}$ dependence.

Magnetic Vector Potential (A). (*MRSE*) Analogous to the electric scalar potential, a magnetic vector potential $\mathbf{A}$ can be defined (Zangwill (2013))

$$\mathbf{A}(\mathbf{r}) = \frac{\mu_0}{4\pi} \int_V \frac{\mathbf{J}(\mathbf{r}_j)}{|\mathbf{r} - \mathbf{r}_j|} dV_j \tag{M.66}$$

where μ_0 is the magnetic permeability of free space and $\mathbf{r}$ is the position vector relative to the origin, $\mathbf{r}_j$ represents the position of the current density $\mathbf{J}(\mathbf{r}_j)$, and the integral is taken over the volume V_j of the source charges. For a simple line current $I(\mathbf{r}_j)$ over a differential length $d\mathbf{l}$ (Griffiths (2017))

$$\mathbf{A}(\mathbf{r}) = \frac{\mu_0}{4\pi} \int_C \frac{\mathbf{I}(\mathbf{r}_j)}{|\mathbf{r} - \mathbf{r}_j|} dl_j = \frac{\mu_0}{4\pi} \int_C \frac{I(\mathbf{r}_j)}{|\mathbf{r} - \mathbf{r}_j|} d\mathbf{l}_j \tag{M.67}$$

The magnetic field may be calculated from the magnetic vector potential $\mathbf{A}$ by (Zangwill (2013))

$$\mathbf{B} = \nabla \times \mathbf{A} \tag{M.68}$$

The electric field may be calculated from the magnetic vector potential $\mathbf{A}$ and electric scalar potential $V(\mathbf{r})$ by (Zangwill (2013))

$$\mathbf{E} = -\frac{\partial \mathbf{A}}{\partial t} - \nabla V(\mathbf{r}) \tag{M.69}$$

Magnetization (M). The process in which individual atomic magnetic dipoles develop coordinated alignment in a preferential direction. The object then has a macroscopic net magnetic moment and produces a net magnetic field of its own. Also may colloquially refer to the magnetic moment. The net magnetization of the object may interact with the magnetic field of the scanner, for example producing an attractive force on the object.

(*MRSE*) The magnetization per unit volume $\mathbf{M}$ is defined by (Coey (2010))

$$\mathbf{M} = \frac{d\mathbf{m}}{dV} \tag{M.70}$$

for magnetic moment $\mathbf{m}$ and volume V, so that

$$\mathbf{m} = \int_V \mathbf{M} dV \tag{M.71}$$

which implies that for a magnetically homogeneous object

$$\mathbf{M} = \frac{\mathbf{m}}{V} \tag{M.72}$$

In response to an external magnetic field $\mathbf{H}_0$, an object becomes magnetized in accordance with its magnetic susceptibility. The object develops an internal field within it $\mathbf{H}_{in}$ which is the sum of the applied $\mathbf{H}_0$ field and a demagnetizing field $\mathbf{H}_d$ which typically opposes the direction of $\mathbf{H}_0$ (Zangwill (2013), Brown et al. (2014)). See Demagnetizing Field.

Using the relationships

$$\mathbf{B}_0 = \mu_0 \mathbf{H}_0 \tag{M.73}$$

and

$$\mathbf{M} = \boldsymbol{\chi}_e \mathbf{H}_0 \tag{M.74}$$

as well as

$$\boldsymbol{\chi}_e = \mathbf{R}\boldsymbol{\chi}_p \mathbf{R}^{-1} \tag{M.75}$$

the magnetization $\mathbf{M}$ is related to the magnetic flux density by (Schenck (2000) Schenck and Kangarlu (2014))

$$\mathbf{M} = \boldsymbol{\chi}_e \mathbf{H}_0 = \mathbf{R}\boldsymbol{\chi}_p \mathbf{R}^{-1} \mathbf{H}_0 = \mathbf{R}\boldsymbol{\chi}_p \mathbf{R}^{-1} \frac{\mathbf{B}_0}{\mu_0} \tag{M.76}$$

where $\mathbf{B}_0$ is the magnetic flux density, $\mathbf{H}_0$ is the applied magnetic field, $\boldsymbol{\chi}_e$ is the effective magnetic susceptibility tensor in the scanner frame, $\boldsymbol{\chi}_p$ is the magnetic susceptibility tensor in its principle axis system, $\mathbf{R}$ is the Euler rotation matrix, and μ_0 is the magnetic permeability of free space. See Magnetic Susceptibility Tensor.

Following the example given by (Schenck (2000) and Schenck and Kangarlu (2014)) the components of $\mathbf{M}$ for an ellipsoid with magnetization induced by the static field when the long axis of the ellipsoid, which is the x-axis in the principle axis system of the ellipsoid, is rotated around the y-axis to θ degrees from the scanner (lab) frame ($\beta = \theta - \pi/2$). It is further assumed that $\mathbf{B}_0 = \big(0, 0, B_0(z)\big) = \big(0, 0, B_0\big)$ and magnetic susceptibility $\chi_{xx} = \chi_{yy} = \chi_{zz} = \chi_m$ and demagnetizing factors $N_{xx} = N_c$ and $N_{yy} = N_{zz} = N_a = N_b$.

Note that because the principle axis system of the magnetic susceptibility tensor is not collinear with that of the magnetic field, an appropriate rotation of the magnetic susceptibility tensor in the principle axis system $\boldsymbol{\chi}_p$ must be performed using the relevant Euler rotation matrix $\mathbf{R}$ (See Magnetic Susceptibility Tensor)

The components of $\mathbf{M}$ can then be calculated as (Schenck (2000), Schenck and Kangarlu (2014))

$$M_z = \frac{\chi_m B_0}{\mu_0}\left[\frac{\cos^2\theta}{1+\chi_m N_c} + \frac{\sin^2\theta}{1+\chi_m N_a}\right] \qquad \text{(M.77)}$$

and

$$M_x = \frac{\chi_m^2 B_0}{\mu_0} \left[\frac{(N_a - N_c)}{(1 + \chi_m N_a)(1 + \chi_m N_c)} \right] \cos\theta \sin\theta \tag{M.78}$$

Note that for alignment of the long axis of the ellipsoid along the z-axis ($\theta = 0$) the magnetization is entirely along the z-axis and given by

$$M_z = \frac{\chi_e B_0}{\mu_0} = \frac{\chi_m B_0}{(1 + \chi_m N_c)\mu_0} \tag{M.79}$$

As the magnetization of an object proceeds where it can become saturated, and even though the field strength increases, the object does not become further magnetized.

Magnetohydrodynamic Effect. Lorentz force from charges in blood moving through the magnetic field. This is the origin of EKG aberrations commonly observed when monitoring patients in an MRI scanner (Chakeres et al. (2003), Schenck and Kangarlu (2014), Gregory et al. (2016)).

(*MRSE*) For conductive tissues in the absence of a substantial external magnetic field the electric current density **J** which gives rise to processes such as the electrical activity of the heart and associated EKG is produced from an associated electric field **E** such that for a conductivity σ_c

$$\mathbf{J} = \sigma_c \mathbf{E} \tag{M.80}$$

If there is motion of the tissue in a magnetic field, this is modified to

$$\mathbf{J} = \sigma_c \left(\mathbf{E} + \mathbf{v} \times \mathbf{B} \right) \tag{M.81}$$

reflecting the motion at a velocity $\mathbf{v}$ through the field $\mathbf{B}$. Note that this effect is most pronounced when the velocity vector is perpendicular to the magnetic field such as blood flowing in the aortic arch.

Magnetophosphenes. Flashes of light seen during an MRI examination caused by induced currents in the visual apparatus caused by Faraday's Law. These are not thought to be dangerous (McRobbie (2014)).

Manufacturer and User Facility Device Experience Database (MAUDE). The FDA MAUDE database is a database of adverse events related to medical devices. There is both mandatory (for example from manufacturers) and voluntary (for example from patients) reporting of adverse events (Clausen et al. (2018), MAUDE (2020), Delfino et al. (2019)).

Matrix. An array of numbers. In MRI data points are collected into a matrix prior to postprocessing and Fourier transformation to yield an image (Brown et al. (2014)).

MAUDE Database. See Manufacturer and User Facility Device Experience Database.

Maximum Effective Slew Rate. Measure of the slew rate if all three x, y, and z gradients are activated simultaneously to their maximum strength.

(*MRSE*) Vector sum of the three maximum slew rates per axis. For equal gradients the maximum effective slew rate $SR_{Max_{Eff}}$ is

$$SR_{Max_{Eff}} = \frac{1}{t_r}\sqrt{G_{z_x}{}^2{}_{Max} + G_{z_y}{}^2{}_{Max} + G_{z_x}{}^2{}_{Max}}$$

$$= \frac{\sqrt{3G^2_{Max_A}}}{t_r} = \frac{G_{Max_A}\sqrt{3}}{t_r} = \sqrt{3}(SR_{Max_A}) \tag{M.82}$$

where t_r is the minimum rise time and the $G_{z_i Max}$ are the maximum gradient strengths in the $i = x, y, z$ directions, which may all have the same maximum amplitude G_{Max_A}. SR_{Max_A} is the maximum slew rate per axis.

Maximum Slew Rate Per Axis. Maximum slew rate for a given gradient axis. For example, assuming all of the maximum gradient amplitudes are equal, the maximum slew rate per axis (SR_{Max_A}) is

$$SR_{Max_A} = \frac{G_{Max_A}}{t_r} \tag{M.83}$$

where G_{Max_A} is the maximum gradient amplitude per axis and t_r is the minimum rise time.

Maximum Spatial Gradient. The point in space where the magnitude of the spatial gradient is greatest, usually near the face of the bore. This is one of the determinants of the magnitude of the magnetic displacement force on an object in the MR environment and is a frequent safety parameter found on the labeling of Conditional objects (Shellock et al. (2011), ASTM F2052-15 (2015)).

The maximum value of the spatial gradient is typically located in the area just inside the entrance of the bore. The value of the maximum spatial gradient along the bore can be approximated as $(dB_0/dz)_{Max}$. For an unsaturated object the maximum displacement force is proportional to the maximum force product of the field and the spatial gradient, while for a saturated object the maximum force has a field dependence only depending on the maximum spatial gradient. The value of the maximum spatial gradient given by a vendor could potentially be in an inaccessible area deep to the magnet cover.

Example: The labeling of a Conditional implant indicates a limit for the maximum spatial gradient of 2000 G/cm. The MRI scanner lists a maximum spatial gradient of 14 T/m. Is the object safe to scan? Noting that 1 T/m = 100 G/cm, the maximum spatial gradient of the scanner is 1400 G/cm, which is less than 2000 G/cm, so that the displacement force on the object should not be of concern for this particular scanner.

(*MRSE*) More generally, the magnitude of the maximum spatial gradient is given by $|\nabla B_0|_{Max}$, which for $\mathbf{B}_0 = \big(0, 0, B_0(z)\big) = \big(0, 0, B_0\big)$, implies

$$|\nabla B_0|_{Max} = \left(\frac{dB_{0_z}}{dz}\right)_{Max} \qquad \text{(M.84)}$$

See Magnetic Displacement Force.

Maximum Static Field Gradient. See Maximum Spatial Gradient.

Maximum Time Varying Gradient Strength. The maximum value of the time varying gradient strength, around 10-80 mT/m (Liang and Lauterbur (2000), Ansorge and Graves (2016)). Not to be confused with the maximum static gradient, which is on the order of 1-40 T/m (Shellock et al. (2011), ASTM F2052-15 (2015)).

Maxwell Coil. Paired set of gradient coils used to produce a time varying gradient along the z-axis in the direction of the main magnetic field (Brown et al. (2014)).

Maxwell's Equations. (*MRSE*) A set of four equations which fully provide a classical description of electromagnetic radiation. Each of these may be expressed in differential or integrated form. They include Gauss's Law for Electricity, Gauss's Law for Magnetism, Faraday's Law, and Ampere's Law (Zangwill (2013)). See individual entries for each law.

Mechanical Safety. Care must be taken to ensure objects or body parts do not become trapped between mechanical parts of the scanner such as the table, which may result in injury (Delfino et al. (2019)).

Medical Emergencies. For medical emergencies occurring in Zone IV, basic life support to stabilize the patient should be initiated by MR personnel while the patient is moved to a magnetically safe location (ACR Committee on MR Safety (2020)).

Medical Waste. See Infection Control.

Medicare Reimbursement in Accordance with the Medicare Improvements for Patients and Providers Act of 2008 (MIPPA). (*MRSE*) Act requiring suppliers of the technical component of advanced diagnostic imaging services be accredited by a designated accrediting organization in order to receive Medicare reimbursement (MIPPA (2008)).

MedWatch. The Safety Information and Adverse Event Reporting Program administered by the US Food and Drug Administration. Adverse advents related to MRI scanning would be reported through this program (Delfino et al. (2019), ACR Committee on MR Safety (2020)).

Metal Detector. A device which detects the presence of metal but does not distinguish ferromagnetic metals such as iron from nonferromagnetic metals such as copper. Metal detectors, as opposed to ferromagnetic detectors, are not recommended as an adjunct to screening by the ACR (ACR Committee on MR Safety (2020)).

Minimal Sedation. (*MRMD*) State of sedation in which the patient can respond normally to verbal commands (Reinking (2014)).

MIPPA. See Medicare Reimbursement in Accordance with the Medicare Improvements for Patients and Providers Act of 2008.

Missile Effect. The effect of a magnetic object being pulled forcefully at high speed into the magnet like a missile. This results from the magnetic displacement force and has been responsible for injuries and deaths involving patients and personnel as well as costly damage to equipment (Keene (2014)).

Moderate Sedation. Sedation level in which patients respond purposefully to verbal commands either alone or with light tactile stimulation (Reinking (2014)).

Monitoring Equipment. There are many situations in which patients require hemodynamic monitoring while undergoing MRI. As the monitoring leads are composed of conductive material, these have the potential to have undesirable currents generated within them which can potentially cause heating and burns. For this reason only MR Conditional monitoring equipment should be used with strict adherence to the instructions for use. As a general rule, it is important to avoid looping the leads, although the manufacturer's guidelines must be followed for safe use (ACR Committee on MR Safety (2020)).

MR. Magnetic Resonance.

MR Attending Physician. The attending physician responsible for a given MRI examination (ACR Committee on MR Safety (2020)).

MR Compatible. Old term no longer used. There is no such thing as an MR Compatible object or device (ACR Committee on MR Safety (2020)).

MR Conditional. An object determined to present no known hazards in a specified MR environment and specified conditions of use. MR Conditional objects may be classified as Conditional1-8 (Shellock (2020)).

MR Conditional1. An object which is weakly affected by the MR environment, but the effects are not clinically significant when the device is used as intended. In these cases, physiological mechanical forces on the object may be greater than the magnetic forces on the object (Shellock (2020)).

MR Conditional2. Weakly ferromagnetic implants such as vascular stents which become incorporated into tissue after six weeks (Shellock (2020)).

MR Conditional3. Transdermal patches with metallic backing which may heat up during MRI (Shellock (2020)).

MR Conditional4. Halo vests and cervical fixation devices with undetermined field interactions (Shellock (2020)).

MR Conditional5. Object which is acceptable for MRI if specific conditions for scanning specified by the manufacturer with respect to the object are followed (Shellock (2020)).

MR Conditional6. An object may be scanned immediately after placement under the following conditions: Static magnetic field $\leq$ 3 T, maximum spatial gradient of 720 G/cm, and maximum WB SAR 2 W/kg for 15 min of scanning per pulse sequence. Temperature statement: In nonclinical testing there was $\leq$ 6 °C at 3 T and 2 W/kg WB SAR. Artifact statement: Image quality may be affected at or near the implant (Shellock (2020)).

MR Conditional7. Object which may be used in the MR environment but not intended for use during active scanning (Shellock (2020)).

MR Conditional8. An object may be scanned immediately after placement under the following conditions: Static magnetic field 1.5 T or 3 T, maximum spatial gradient of 720 G/cm, and maximum WB SAR 2 W/kg for 15 min of scanning per pulse sequence. Temperature statement: In nonclinical testing there was $\leq$ 6 °C at 1.5 T and 3 T and 2 W/kg WB SAR. Artifact statement: Image quality may be affected at or near the implant (Shellock (2020)).

MR Environment. As defined by the ASTM, the area with magnetic flux density of greater than 5 G (0.5 mT) (ASTM F2213-17 (2017)).

MRI. Magnetic Resonance Imaging.

MRMD. See MR Medical Director.

MR Medical Director (MRMD). One of the certifications of the ABMRS and a formal safety role recommended by the ACR. A licensed physician with appropriate training in MR safety with overall and ultimate responsibility for operational safety of an MR facility (Calamante et al. (2016), ACR Committee on MR Safety (2020)). The MRMD ensures at all times, either in person or by delegation to another qualified individual, the satisfactory performance of the safe execution of all MR examinations, the appointment of an MRSO and advisory MRSE, the development, implementation, and maintenance of specific policies and procedures for safe MR services, the implementation and maintenance of appropriate MR safety and quality assurance programs, the development of an appropriate system for record keeping and analysis of adverse events with input from the MRSO and MRSE as needed), the development of an appropriate ongoing assessment of risk for the facility, and development of an appropriate investigation and recording of all reported MR safety adverse events. The MRMD identifies the necessary training for and also identifies individuals qualifying as Level 1 and Level 2 MR Personnel.

MR Personnel. Individuals documented have successfully completed at least annual education regarding MR safety issues to at least a level as defined by the site's MRMD that they do not pose a danger to themselves or others in the MR environment and at a greater level than what might be provided to Non-MR Personnel (ACR Committee on MR Safety (2020)).

MR Physicist. A qualified medical physicist is an individual competent to provide independent clinical professional services in magnetic resonance imaging physics, has a graduate degree in a relevant field, and has certification in the subfield of MRI physics (ACR (2018)).

MDRD Equation. See Glomerular Filtration Rate.

MR Safe. An object which presents no known hazards in any MR environment. Example: Completely plastic ID badge (Shellock (2020)).

MR Safety Certified (MRSC™). On achieving one of the ABMRS certifications an individual is then considered MR Safety Certified (ABMRS).

MRI Safety Committee. A committee responsible for overseeing and implementing MRI safety practices at an MRI site such as a hospital, imaging center, or research facility. May be composed of an MRSO, MRMD, MRSE, and representatives from administration and representatives from nonimaging staff such as security and housekeeping.

MR Safety Expert (MRSE). One of the certifications of the ABMRS and a formal safety role recommended by the ACR. An advisory position to the MRSO and MRMD with substantial training on MR safety and regulatory issues with greater understanding of quantitative aspects of MR safety. Usually but not always an MRI physicist (Calamante et al. (2016)).

The MRSE may be external to the organization and serve one or more MRI sites and may not be physically onsite. The MRSE should be prospectively identified by each site with a prospectively identified and clear method of contact. MRSE responsibilities include providing advice on engineering, scientific, and administrative aspects of MR safety, including quantification analysis of energy, force, and risk exposures, providing advice on the development and continuing evaluation of a safety framework for the site's MRI environment, providing advice for the development of local policies and procedures to ensure the safe use of MRI equipment, providing safety advice regarding nonroutine MRI procedures, including advice regarding safety related to implanted devices and other similar issues; providing advice on the choice of MR Safety programs and MR Quality Assurance programs, evaluations, and audits, providing safety advice regarding equipment acceptance testing, establishing and maintaining links with appropriate regional and professional bodies and reporting back to the MRMD and MRSO on safety-related issues, and providing expertise in root cause analyses, solutions meetings, etc, related to adverse MRI events (ACR Committee on MR Safety (2020)).

MR Safety Officer (MRSO). One of the certifications of the ABMRS and formal safety role recommended by the ACR. The MRSO is responsible for the implementation of MR safety procedures and policies in the department under the direction of the MR Medical Director. This individual is usually an MR technologist (Calamante et al. (2016)).

Multiple MRSOs may be present at a site, but a single MRSO should be identified as responsible for and overseeing safety practices within a defined component of the MRI practice at all times.

MRSO responsibilities include ensuring access at all times to the operators of active MRI facilities, ensuring proper policies and procedures of the MRMD are implemented and enforced, ensuring development, documentation, and execution of safe working procedures for the MR environment together and under the authority of the MRMD, ensuring adequate written safety, emergency procedures, and operating procedures are issued, in consultation with the MRMD and MRSE when appropriate, ensuring implementation and monitoring of measures to minimize risks to staff and patients, in cooperation with the MRMD, managing hazards posed by the MRI scanner and monitoring the measures taken to protect against those hazards, ensuring together with the MRMD, that medical, technical, nursing, emergency, and all other relevant staff groups (including ancillary workers) who may be exposed to the MR environment are educated appropriately and updated as necessary as to MR safety requirements, providing and/or ensuring the provision of MR safety education and training in cooperation with and per the policies of the MRMD and maintaining records of personnel education, consulting the MRMD and/or MRSE when further advice is required regarding MR safety, reporting back to the MRMD in a timely fashion any and all MR safety-related issues, and ensuring that there is a clear policy for purchasing, testing, and clearly marking of all equipment that will be taken into Zones III and IV, and providing safety advice on the modification of MR protocols in cooperation with the MRMD and/or MRSE as needed, maintaining regular contact with other relevant groups or committees responsible for the safety and welfare of personnel on site, providing expertise in root cause analyses, solutions meetings, etc, related to MRI adverse events.

MRSC™. See MR Safety Certified.

MR Scientist. Requirements include a graduate degree in magnetic resonance and three years of relevant clinical experience (ACR (2018)).

MRSE. See MR Safety Expert.

MRSO. See MR Safety Officer.

MR Technologist. MR technologists should be individuals with the qualifications given by the ACR

MRI Accreditation Program (ACR Committee on MR Safety (2020)). To be ACR accredited a facility's technologists must be state or other jurisdiction licensed if such licensure exists, be registered as an MRI technologist by one of the registries including the American Registry of Radiologic Technologists, The American Registry of Magnetic Resonance Imaging Technologists, or Canadian Association of Medical Radiation Technologists, or have continuously performed MRI since 1996 and been evaluated for competence by a responsible physician (Clausen et al. (2018)). MR Technologists represent perhaps the most important individual in the chain of safety and are typically the last barrier to an MR safety accident.

MR Unsafe. An object which is known to cause a hazard in all MR environments. These can be classified as Unsafe1 and Unsafe2 (Shellock (2020)).

MR Unsafe1. An object which is unsafe in the MR environment due to forces on the object (Shellock (2020)).

MR Unsafe2. An object which is unsafe in the MR environment due to currents induced in the object (Shellock (2020)).

N

National Electrical Manufacturers Association (NEMA). Association of companies which make electrical and medical systems. Dedicated to safe, reliable, and efficient products and systems. Relevant to MR safety, the association has published standards on measurement of acoustic noise (NEMA MS 4-2010), SAR (NEMA MS 8-2016), local SAR (NEMA MS 10-2010), electric fields from gradients (NEMA MS 11-2010), and dB/dt (NEMA MS 7-1993).

National Institute for Occupational Safety and Health (NIOSH). (*MRSE*) An agency associated with the Centers for Disease Control (CDC) dedicated to developing knowledge in occupational safety and health and transitioning that into practice.

Near Field. Area near the rf transmit coil with the capacity for high energy deposition.

(*MRSE*) Near the rf transmit coil there are high inductive and capacitive effects as described below.

Using a model of a short dipole of length l with an alternating current (Hertzian dipole) of frequency ω and maximum amplitude I_0 oriented along the z-axis assuming a current of the form

$$I\left(t\right) = I_0 \cos \omega t \tag{N.1}$$

it can be shown that the electric field **E** and magnetic field **B** around the dipole are given in spherical coordinates by (Rao (2009))

$$\mathbf{E} = \frac{2I_0 l\cos\theta}{4\pi\varepsilon\omega}\left[\frac{\sin(\omega t - kr)}{r^3} + \frac{k\cos(\omega t - kr)}{r^2}\right]\widehat{\mathbf{r}} + \frac{2I_0 l\sin\theta}{4\pi\varepsilon\omega}\left[\frac{\sin(\omega t - kr)}{r^3} + \frac{k\cos(\omega t - kr)}{r^2} - \frac{k^2\sin(\omega t - kr)}{r}\right]\widehat{\boldsymbol{\theta}} \tag{N.2}$$

and

$$\mathbf{B} = \frac{\mu I_0 l\sin\theta}{4\pi}\left[\frac{\cos(\omega t - kr)}{r^2} - \frac{k\sin(\omega t - kr)}{r}\right]\widehat{\boldsymbol{\phi}} \tag{N.3}$$

where $k = 2\pi/\lambda$, ε is the electric permittivity, and μ is the magnetic permeability. These waves are spherical waves. At short distances there are multiple components of the fields in what is termed the near field region. The fields decay away in terms of $1/r^3$, $1/r^2$, and $1/r$. For the area close to the dipole $kr << 1$ the fields are dominated by the $1/r^3$ term for **E** and the $1/r^2$ for **B** so that

$$\mathbf{E} = \frac{2I_0 l\cos\theta}{4\pi\varepsilon\omega}\left[\frac{\sin(\omega t - kr)}{r^3}\right]\widehat{\mathbf{r}} + \frac{2I_0 l\sin\theta}{4\pi\varepsilon\omega}\left[\frac{\sin(\omega t - kr)}{r^3}\right]\widehat{\boldsymbol{\theta}} \tag{N.4}$$

and

$$\mathbf{B} = \frac{\mu I_0 l \sin\theta}{4\pi}\left[\frac{\cos(\omega t - kr)}{r^2}\right]\widehat{\boldsymbol{\phi}} \tag{N.5}$$

These are the same forms as an electric dipole field and magnetic field from a differential current element but modulated by the time dependent trigonometric terms (compare Eqs. B.34 and E.8). Note that as the system oscillates at ωt at some points the system purely stores electric charge and functions as a capacitor while at others the system stores a magnetic field and functions as an inductor. There is the potential for high amounts of energy to be present close to the source of the rf.

Conversely, for $kr >> 1$ the fields are dominated by the $1/r$ term for both **E** and **B** and have the form

$$\mathbf{E} = \frac{I_0 l \sin\theta}{4\pi\varepsilon\omega}\left[-\frac{k^2\sin(\omega t - kr)}{r}\right]\widehat{\boldsymbol{\theta}} \tag{N.6}$$

and

$$\mathbf{B} = \frac{\mu I_0 l \sin\theta}{4\pi}\left[-\frac{k\sin(\omega t - kr)}{r}\right]\widehat{\boldsymbol{\phi}} \tag{N.7}$$

This region is termed the far field and is described by a spherical wave with orthogonal **E** and **B** components whose intensity decays by $1/r^2$. These features are consistent with the known properties of electromagnetic radiation. At sufficient distance the curvature is sufficiently gentle to model the spherical wave as a plane wave.

Near Field Burn. A burn resulting from close proximity of the body to the rf transmit coil. This is usually caused by the body coil. Manufacturers recommend that there be 0.5 to 1 cm between the skin surface and the inner surface of the bore. Padding is provided for this purpose. These burns tend to be small and round on the peripheral portions of the body such as the outer arms (Tsai et al. (2015), Mandel et al. (2017)). These burns are also referred to as proximity burns.

(*MRSE*) A burn is more likely to result when the patient is in the vicinity of the coil where the near field components have substantial contributions to the field. See Near Field.

NEMA. See National Electrical Manufacturers Association.

Nephrogenic Systemic Fibrosis (NSF). First described in 2000 associated with chronic kidney disease, NSF is a fibrosing illness primary affecting the skin and subcutaneous tissues but occasionally affecting the deep soft tissue organs. This may be progressive and lead to contractures or death. Subsequently it was discovered that the disease is associated to prior exposure to gadolinium based contrast agents. It is thought to be mediated by dissociation of the gadolinium ion from its ligand. In the setting of renal failure there is decreased excretion with a prolongation of the time the contrast remains in the body compared to normal individuals. The most widely held hypothesis for the mechanism of NSF involves an initial step of free gadolinium dissociation from its chelate with transmetallation replacement of another ion such as calcium in a complex with an anion such as phosphate to form a compound which precipitates in soft tissues, inciting a fibrotic reaction (ACR Committee on Drugs and Contrast Media (2020) McDonald et al. (2018)).

Risk factors for developing NSF in the setting of gadolinium administration include acute kidney injury and chronic kidney disease. Repeated doses and high doses of gadolinium based contrast are also risk factors.

(*MRMD*) Administration of gadolinium based contrast media (GBCM) to at risk patients with chronic kidney disease (CKD) may result in the development of NSF. Per the ACR Contrast Manual if the potential benefits of gadolinium administration are felt to outweigh the risk of NSF in an individual at risk patient, the referring physician and the patient (if possible) should be informed of the risk, and both should agree to the injection of contrast. Group II gadolinium agents are strongly preferred in at risk patients. The lowest dose to obtain the needed clinical information should be used. For patients on chronic hemodialysis, elective examinations should be timed close to a subsequent dialysis session.

CKD1-5 stages (Simon et al. (2011)) with recommendations for gadolinium administration from the ACR Contrast Manual are given below:

CKD1: GFR $\geq 90\,m\mathrm{L}/min/1.73\,m^2$. No evidence of risk from GBCM.
CKD2: GFR 60-89 $m\mathrm{L}/min/1.73\,m^2$. No evidence of risk from GBCM.
CKD3: GFR 30-59 $m\mathrm{L}/min/1.73\,m^2$. No special

precautions necessary.
CKD4: GFR 15-29 $mL/min/1.73\,m^2$. Group I gadolinium agents are contraindicated in this population group. If gadolinium administration is necessary, a Group II agent should be used.
CKD5: GFR $< 15\,mL/min/1.73\,m^2$. Group I gadolinium agents are contraindicated in this population group. If gadolinium administration is necessary, a Group II agent should be used.

There is an estimated 1-7% chance of developing NSF following administration of one or more doses of a Group I agent. Per the ACR Contrast Manual there is very little, if any, risk of developing NSF from Group II agents, regardless of renal status. There is insufficient data for Group III agents.

Newton (N). (*MRSE*) Unit of force. 4.45 Newton = 1 pound (Halliday et al. (2014)).

NEX. Number of excitations. Also called number of signal averages. This contributes to SED but not to SAR (Brown et al. (2014), Faulkner (2016)).

NIOSH. See National Institute for Occupational Safety and Health.

NIOSH Hierarchy of Controls. (*MRSE*) The hierarchy of controls to mitigate safety risks include elimination, substitution, engineering controls, administrative controls, and personal protective equipment (NIOSH, (2019)).

Noise Reduction Rating (NRR). The noise reduction rating (NRR) describes the sound level attenuation based on laboratory evaluation for a sound attenuation device such as earplugs. To obtain the effect of the attenuation the following equation is used (Occupational Safety and Health Administration Standard 29 CFR 1910.95, McRobbie (2020))

$$\text{Attenuated Noise Level} = \text{Noise Level} - (\text{NRR} - 7)/2 \tag{N.8}$$

Example: For a noise level of 100 dB(A) and earplugs with NRR 29, the attenuated noise level is

$$\text{Attenuated Noise Level} = 100 - (29 - 7)/2 = 89\,\text{dB(A)} \tag{N.9}$$

Nonallergic Reaction. Physiological reaction to contrast not mediated by or clinically resembling typical immune mediated reactions. Nonallergic reactions to gadolinium include dysesthesias at the injection site, headache, nausea, vomiting, dizziness, and paresthesias (ACR Committee on Drugs and Contrast Media (2020)).

Non-MR Personnel. Individuals not meeting yearly training criteria for MR personnel as defined by the MRMD of that particular site. Of note, MR personnel who do not renew their training yearly revert to Non-MR personnel after one year. Each Non-MR Personnel entering Zone III must have a specifically identified Level 2 MR Personnel, usually but not always an MR technologist, responsible for them throughout their stay in Zones III and IV. This function of the Level 2 MR Personnel is directly under the authority and responsibility of the MRMD or the responsible Level 2 attending

MR Physician. The specifically assigned Level 2 personnel must be immediately supervising and in visual contact with the Non-MR Personnel while in Zones III and IV, except when the Non-MR personnel are in the changing room or bathroom, in which case verbal communication is sufficient (ACR Committee on MR Safety (2020)).

Normal Operating Mode. Mode of MRI scanner operation in which the output is not expected to cause physiological stress. This should be the default mode. Quantitative limits for this mode have been published by the IEC (Keevil (2014), Schaefer (2014)). See Exposure Limits.

NSF. See Nephrogenic Systemic Fibrosis.

NRR. See Noise Reduction Rating.

O

Occupational Safety and Health Administration (OSHA). (*MRSE*) An agency of the US Department of Labor which ensures safe and healthy work environments. Regulations which are relevant to MRI safety include an rf exposure limit of $10\,mW/cm^2 = 100\,W/m^2$ and energy density $1\,mW\text{-}hr/cm^2$ during any 6 minute period (Occupational Safety and Health Administration Standard 29 CFR 1910.97). Noise limits are given as a function of hours of exposure per day (Occupational Safety and Health Administration Standard 29 CFR 1910.95). For example, a limit of 90 dB(A) for 8 hours is imposed with a limit of 115 dB(A) for 15 minutes or less. Requirements for use of breathing apparatus is also relevant in the setting of a quench breach in which cryogens are discharged into the scanner room (Occupational Safety and Health Administration Standard 29 CFR 1915.12). Safety signs must be readable at 5 feet (Occupational Safety and Health Administration 29 CFR Standard 1910.145).

Ohm's Law. An expression relating voltage (V), current (I), and resistance (R), $V = IR$ (Halliday et al. (2014)). This is important for heating during MRI.

(*MRSE*) Ohm's Law in general form is (Zangwill (2013)).

$$\mathbf{J} = \sigma_c \mathbf{E} \tag{O.1}$$

where $\mathbf{J}$ is current density, σ_c is conductivity, and $\mathbf{E}$ is the electric field.

The relationship to the macroscopic form of Ohm's Law can be seen by considering a current density flowing along a wire of length l and cross sectional area A. Integrating the electric field $\mathbf{E}$ over the wire gives the voltage V

$$V = \int_C \mathbf{E} \cdot d\mathbf{l} = El \tag{O.2}$$

while integrating the current density $\mathbf{J}$ over the area A assuming that $\mathbf{J}$ flows perpendicular to the cross section of the wire and using the definition of current density as current I per unit area gives

$$I = \int_S \mathbf{J} \cdot d\mathbf{A} = JA \tag{O.3}$$

Substituting into Eq. O.1 gives

$$\frac{I}{A} = \sigma_c \frac{V}{l} \tag{O.4}$$

so that

$$V = \frac{Il}{\sigma_c A} = IR \tag{O.5}$$

with resistance R

$$R = \frac{l}{\sigma_c A} \tag{O.6}$$

For alternating current (AC) conditions the impedance Z is used

$$V = IZ \tag{O.7}$$

Ohmic Heating. See Joule Heating.

Ohmic Power Density. See Power Density.

Organizations. Organizations relevant to MR safety in the US are the American Board of Magnetic Resonance Safety (ABMRS), American College of Radiology (ACR), American Society of Radiologic Technologists (ASRT), American Society for Testing and Materials International (ASTM), Food and Drug Administration (FDA), Heart Rhythm Society (HRS), National Electrical Manufacturers Association (NEMA), National Institute for Occupational Safety and Health (NIOSH), International Commission on Non-Ionizing Radiation Protection (ICNIRP), International Electrotechnical Commission (IEC), Intersocietal Accreditation Commission (IAC), Occupational Safety and Health Administration (OSHA), International Organization for Standardization (ISO), RadSite, and The Joint Commission (TJC).

OSHA. See Occupational Safety and Health Administration.

P

Pacemaker. An active implant used to regulate heart rhythm. Some pacemakers are labeled Unsafe and some Conditional. Pacemakers which are labeled unsafe may be considered for MRI provided certain protocols are used (Indik et al. (2017)).

Potential problems which may arise depending on the model, programming, and MRI conditions include loss of data, reset of the device, change in programming, loss of capture threshold, myocardial heat damage, battery damage, inhibition of pacing, battery depletion, forces which could damage the device, and induced arrhythmia (Indik et al. (2017), ACR Committee on MR Safety (2020)).

Paramagnetic. Materials in which the atomic dipoles are randomly oriented, but become partially aligned in an external magnetic field to produce a magnetized material, but the alignment disappears in the absence of the external field. Paramagnetic materials are characterized by small positive relative permeabilities and magnetic susceptibilities with susceptibilities on the order of 10^{-3} to 10^{-5} Example: Aluminum (Jiles (2015)). See Magnetic Susceptibility.

Parolees. See Prisoners and Parolees.

Passive Implant. Medical implant which does not use electrical power (ASTM F2182-19e2 (2019)). Example: Orthopedic plate and screw fixation.

Passive Shielding. Shielding of the magnetic field built into the walls of Zone IV. This is made of silicon steel and acts as a guide for the magnetic field within the wall, which functions to contain the field. Passive shielding is usually avoided because of its weight and expense to install, and active shielding is commonly used (Junk (2014)). See Active Shielding.

Patient Heating. Global patient heating occurs through magnetic induction from the applied rf field. Patient heating is decreased through cooling mechanisms. See Cooling. Maneuvers to decrease patient heating are use of GRE, SSGRE, or EPI as opposed to SE or FSE, avoid fat saturation pulses, avoid magnetization transfer, avoid inversion recovery, decrease flip angle, increase T_R, increase TE, decrease slices in a given T_R, increase concatenations, increase pulse duration, control room temperature and humidity, interleave high and low SAR acquisitions, use bore ventilation fan, and avoid heavy blankets. Note that larger patients will have intrinsically greater magnetic induction (increased radius) and intrinsically decreased cooling (decreased surface area to volume) (Brown et al. (2014), Schaefer (2014), Allison and Yanasak (2015), (Shellock (2020)). Some patients have impaired thermoregulation such as patients with decreased cardiac function, hypertension, advanced age, obesity, fever, impaired perspiration, pregnancy, and some pharmaceutical agents (Allison and Yanasak (2015)).

Patient Height. Patient height figures into MRI safety to the extent that it determines what portions of the patient's anatomy may be in the regions of the magnet relevant to the safety issues regarding a particular object in the patient.

An example is shown in Figure P.1. For simplicity, it is assumed that the patient is 180 *cm* tall, with sections of the body as shown. The scanner bore is assumed to be 120 *cm*. In one case the patient's head and torso are in the scanner with legs and feet outside, and in the second case the patient's legs and feet are in the scanner with the head and torso outside. Suppose there is an implant in the region of the patient's groin. In both cases the implant may be just at the edge of the bore, and therefore in a region of low irradiation from both the applied rf and time varying gradient fields. However, this is in the area near the maximum spatial gradient of the static field where the magnetic displacement force would be maximal on a magnetic object. If, however, the patient's head is being scanned, and there is for example a wire-like metallic object in the leg, it might be reasonable to conclude, after examining the field plots for the scanner, that the benefits of scanning outweigh the risks based on the location of the object in the body relative to the scanner components.

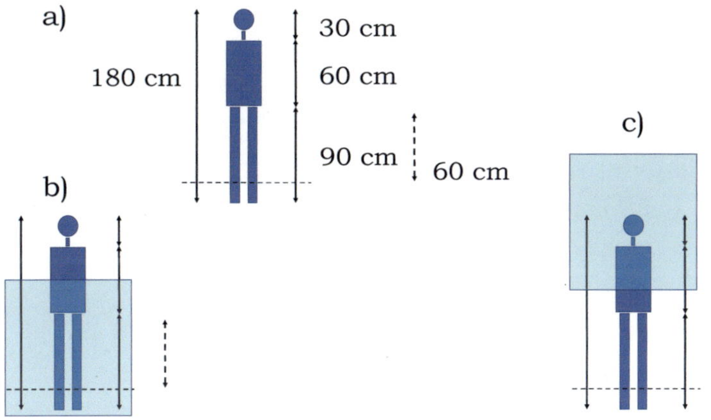

Figure P.1: Dimensions of a 180 cm patient as shown in (a). Same patient with legs in scanner (b). Same patient with head in scanner (c).

Patient Monitoring. See Monitoring Equipment.

Pediatrics. Younger pediatric patients often require sedation. Adherence to the appropriate guidelines for sedation are required. See Anesthesia. In addition, in the neonatal and young patient population, extra attention to thermal homeostasis and other vital signs may be necessary. In terms of screening, the reliability of histories obtained from children and adolescents must be considered, and it may be appropriate to question adolescents in particular both in the presence and absence of parents. It is also necessary to remember that transitional objects such as blankets or toys could carry a safety risk in the MR environment (ACR Committee on MR Safety (2020)).

Pennes Bioheat Equation. (*MRSE*) A mathematical model of heat flow in biological tissues depending on external heating, internal metabolic heating, and heat losses through blood flow (Carluccio et al. (2016), Shrivastava (2018)). The Pennes Bioheat Equation and others like it are used to model *in vivo* heating from MRI. The Pennes Bioheat Equation is

$$\rho C_p \frac{\partial T}{\partial t} = \nabla\left(\kappa \nabla T\right) - b\left(T - T_b\right) + Q + \rho \mathrm{SAR} \quad \text{(P.1)}$$

where ρ is the density, κ is the thermal conductivity, C_p is the heat capacity, T is temperature, t is time, T_b is blood temperature, Q is heat generated through metabolism, and SAR is the specific absorption rate. b depends on tissue perfusion as well as blood density and blood heat capacity.

For phantom measurements such as are required in ASTM F2182-19e2 (ASTM F2182-19e2 (2019)), where there are no perfusion and metabolic effects and for short time periods during which there are limited conduction effects

$$\frac{\Delta T}{\Delta t} = \frac{\mathrm{SAR}}{C_p} \quad \text{(P.2)}$$

Therefore a plot of measured temperature vs. time can be used to determine SAR if the heat capacity of the medium is known.

Peripheral Nerve Stimulation (PNS). Currents via magnetic induction cause peripheral nerves to fire inducing potentially unpleasant sensations during an MRI. These tend to occur during sequences with rapid gradient switching such as those which use EPI such as diffusion and perfusion imaging (McRobbie (2014)).

(*MRSE*) Peripheral nerve stimulation can be understood in terms of the strength duration curve related to depolarization thresholds. See Strength Duration Curve. Considering the Weiss form of the curve, IEC limits for PNS are defined in terms of the median perception threshold for a current I_{th}, using a rheobase of $I_r = 20\,\mathrm{T}/s$ and chronaxie of $C_h = 0.36\,ms$. The Weiss form is

$$
\begin{aligned}
I_{th}(\tau) &= I_r \left(1 + \frac{C_h}{\tau}\right) \\
&= \left(20\,Ts^{-1}\right)\left(1 + \frac{0.36\,ms}{\tau}\right)
\end{aligned}
\tag{P.3}
$$

where τ is the stimulus duration. The Normal Operating Mode limit (L01) is then defined as 80% of the median perception threshold while the First Level Controlled Operating Mode (L12) is defined as 100% of the median perception threshold (McRobbie (2014)).

Example: Calculate the Normal Operating Mode and First Level Controlled Operating Mode limits for a stimulus of $0.5\,ms$. Using Eq. P.3

$$
\mathrm{L01} = (0.8)\left(20\,Ts^{-1}\right)\left(1 + \frac{0.36\,ms}{0.5\,ms}\right) = 27.5\,\mathrm{T}s^{-1}
\tag{P.4}
$$

and for First Level Controlled Operating Mode

$$\text{L12} = (1.0)\left(20\,Ts^{-1}\right)\left(1+\frac{0.36\,ms}{0.5\,ms}\right) = 34.4\,\text{T}s^{-1} \tag{P.5}$$

Permittivity of Free Space (ε_0). Constant relating electric fields in vacuum to electric fields in matter. $\varepsilon_0 = 8.854 \times 10^{-12}\ F/m$ (Zangwill (2013)).

Phase Encoding. Constant time application of varying gradient strengths up to a maximum strength determined by the field of view during an MRI sequence to produce an image (Brown et al. (2014)).

Phasor. (*MRSE*) Notation for waves in exponential form without time dependence (Rao (2009)). Considering a time varying wave with amplitude A_0, angular frequency ω, and phase α along the x-axis

$$\begin{aligned}\mathbf{A}(t) &= A_0\cos(\omega t+\alpha)\widehat{\mathbf{x}} = Re\left[A_0 e^{i(\omega t+\alpha)}\right]\widehat{\mathbf{x}} \\ &\Longrightarrow A_0 e^{i\alpha}\end{aligned} \tag{P.6}$$

where $A_0e^{i\alpha}$ is $\mathbf{A}(t)$ written in phasor notation denoted

$$\widetilde{A}(t) = A_0 e^{i\alpha} \tag{P.7}$$

Plane Wave. (*MRSE*) A set of waves that can be considered as a set of planes moving in a given direction. Electromagnetic radiation, such as rf radiation, under many conditions can be modeled as plane waves in which an orthogonal electric and magnetic field move along a third orthogonal direction, varying in space and time (Rao (2009), Zangwill (2013)). See Electromagnetic Waves.

PNS. See Peripheral Nerve Stimulation.

Power (P). Energy per time. Measured in Watts. $1\,W = 1\,J/s$ (Halliday et al. (2014)).

Power Density. (*MRSE*) The deposited power per unit volume (volume power density) or per unit area (surface power density). In MRI safety a quantity of interest is the time averaged ohmic power density (Brown et al. (2014)). The Joule Heating Law gives the power P from ohmic heating in terms of the electric field $\mathbf{E}$ and current density $\mathbf{J}$ as

$$P = \int_V \mathbf{E} \cdot \mathbf{J} dV = \int_V \sigma_c \mathbf{E} \cdot \mathbf{E} dV = \int_V \sigma_c \left|\mathbf{E}\right|^2 dV \tag{P.8}$$

where Ohm's Law $\mathbf{J} = \sigma_c \mathbf{E}$ has been used. σ_c is the conductivity.

The ohmic power density, the change in power with volume, is

$$\frac{dP}{dV} = \sigma_c \left|\mathbf{E}\right|^2 \tag{P.9}$$

The amount of power deposited per volume of tissue averaged over time, the time averaged ohmic power density, is

$$\frac{dP_t}{dV} = \overline{\sigma_c \left|\mathbf{E}\right|^2} = \frac{1}{T} \int_0^T \sigma_c \left|\mathbf{E}\right|^2 dt \tag{P.10}$$

where dP_t/dV denotes time averaging of dP/dV. Considering $\mathbf{E}$ as a time varying function such as (See Electromagnetic Wave)

$$\mathbf{E} = E_A cos\left(kz - \omega t\right)\widehat{\mathbf{x}} \tag{P.11}$$

and assuming a sufficiently long averaging time T, the time averaged ohmic power density is

$$\begin{aligned}
\frac{dP_t}{dV} &= \lim_{T\to\infty}\frac{1}{T}\int_0^T \sigma_c \mathbf{E}\cdot\mathbf{E}dt \\
&= \lim_{T\to\infty}\frac{1}{T}\int_0^T \sigma_c\left|E_A cos\left(kz-\omega t\right)\widehat{\mathbf{x}}\right|^2 dt \\
&= \lim_{T\to\infty}\sigma_c E_A^2\left(\frac{2T\omega + sin(2kz) + sin(2T\omega - 2kz)}{4T\omega}\right) \\
&= \frac{1}{2}\sigma_c {E_A}^2
\end{aligned} \tag{P.12}$$

This is is used to calculate the specific absorption rate. See Specific Absorption Rate.

Pregnancy and Gadolinium Administration. See Gadolinium Contrast in Pregnancy.

Pregnancy and MRI in Patients. Per the 2020 ACR Manual on MR Safety The vast majority of data has not shown deleterious effects exposure of the developing fetus to MRI. Nonetheless, in pregnant patients, the decision to perform a noncontrasted MRI at 1.5 T should be based on the medical benefits weighed against unknown potential risk. In addition, 3 T MR examinations performed within normal operating mode for durations less than 30 minutes have not shown harm in pregnant patients. Ultimately, the decision to image a pregnant patient at 3 T should be based on local institutional policies, medical needs, and accessibility to 1.5 T versus 3 T MR scanners. The safety of imaging pregnant patients at field strengths greater than 3 T, for example 7 T, is unclear (ACR Committee on MR Safety (2020)).

Pregnancy and MRI in Technologists. The 2020 ACR Manual on MR Safety recommendation is that technologists can continue working throughout pregnancy but should not be in the scanner room (Zone IV) during active scanning. This is based on data at 3 T. The safety at higher fields such as 7 T has not been established. (ACR Committee on MR Safety (2020)).

Premedication for Gadolinium Contrast. There are no studies available to confirm the efficacy of premedication to prevent gadolinium contrast reactions (ACR Committee on Drugs and Contrast Media (2020)). The premedication regimens commonly used for iodinated contrast are often used and involve administration of steroids and diphenhydramine (ACR Committee on Drugs and Contrast Media (2020)).

(*MRMD*) The premedication regimens from the ACR Manual on Contrast commonly used for iodinated contrast are shown below (ACR Committee on Drugs and Contrast Media (2020)):

Elective: 50 mg prednisone PO 13 hours, 7 hours, and 1 hour prior to contrast with 50 mg diphenhydramine IV, IM, or PO 1 hour prior to contrast OR

32 mg methylprednisolone PO 12 hours and 2 hours prior to contrast with 50 mg diphenhydramine IV, IM, or PO 1 hour prior to contrast. If the patient cannot take PO, 200 mg hydrocortisone may be substituted for the prednisone.

Urgent: 40 mg methylprednisolone IV immediately then every 4 hours until contrast with 50 mg diphenhydramine IV one hour prior to contrast OR 200 mg hydrocortisone IV immediately then every 4 hours until contrast with 50 mg diphenhydramine IV one hour prior to contrast OR 7.5 mg dexamethasone IV immediately then every 4 hours until contrast with 50 mg diphenhydramine IV one hour prior to contrast.

Emergent: 40 mg methylprednisolone IV and 50 mg diphenhydramine IV one hour prior to contrast OR 200 mg hydrocortisone IV and 50 mg diphenhydramine IV one hour prior to contrast.

Note that premedication regimens of less than 4-5 hours have not been demonstrated to be effective (ACR Committee on Drugs and Contrast Media (2020)).

Pressure (P). Force per unit area. Example pounds per square inch. Pressure is important in MRI safety because in the event of a quench, pressure may build up and cause an explosion or pressure entrapment of those inside the scanner room or potentially rupture eardrums (Halliday et al. (2014), ACR Committee on MR Safety (2020)). Sound pressure level is important in the setting of acoustic noise from scanning and associated potential hearing loss (McJury (2014)).

Pressure Entrapment. During a quench breach, the scanner room may fill with gas from the boiled off cryogens causing pressure build up in the scanner room. If the door to the scanner opens into the room the door may be forcefully held shut from the inside room pressure, trapping anyone in the scanner room inside (ACR Committee on MR Safety (2020)).

Prisoners and Parolees. Prior to the patient arriving in the MRI department, advanced request should be made to the appropriate authorities for a nonferrous restraining option. Screening of the patient and the accompanying correctional personnel then occurs prior to entering Zone III. The accompanying correctional personnel should be educated regarding static magnetic field safety issues, and if they are agreeable, should accompany the technologist into Zone IV for patient positioning and retrieval.

Patients on parole may be wearing identification or tracking bracelets which could be adversely affected in the MR environment related to forces and/or induced currents, both in terms of injury and malfunction. In cases in which one of these patients may be wearing an active or passive restraint device with potentially hazardous components, the patient should be accompanied by appropriate authorities who will remove the restraining device prior to the MR study and replace it following the examination.

As any accompanying security personnel may be carrying firearms, appropriate firearm safety must be observed. See Firearms.

Proximity Burn. See Near Field Burn.

Psychological Distress in MRI. Patients may have a variety of psychological reactions to MRI ranging from mild anxious feelings to severe panic attacks. Feelings of claustrophobia in the magnet bore may contribute. Mild anxiety may not interfere with the patient's ability to undergo MRI, but in severe form may require anesthesia.

The most important interventions for treating the anxious patient are preparation, education, and reassurance prior to the procedure. Allowing a friend or relative to provide support after appropriate screening may also be useful. Music and eye covering or videos are often helpful. Positioning including feet first or prone positioning if possible. Brightly lit scanners and rooms with adequate fan ventilation in the bore, and aroma therapy may be helpful. Relaxation techniques such as controlled breathing, systematic desensitization, hypnosis, or pharmacologic interventions may be useful in cases refractory to the above (Spaeth and Gollub (2014)).

Pulse Length (τ_p). Period of time the exciting rf pulse lasts (Brown et al. (2014)).

Pulse Sequence. A collection of radiofrequency pulses and gradients that are turned on in off in a specific sequence in time to create an MR image. Some sequences are associated with higher SAR than others, and some sequences are more likely to cause peripheral nerve stimulation than others (Brown et al. (2014)).

Pulse Shape. Many pulses in MRI can be modeled as sinc pulses, but different pulse shapes can be produced with different properties. For the purposes of SAR, there is a numeric shape factor which weights the pulse in terms of its power deposition (Schaefer (2014)).

Pulse Time. See Pulse Length.

Pulse Width. See Pulse Length.

Q

QRS Complex. Portion of the EKG corresponding to ventricular depolarization (Faulkner (2016)).

Quadrature Coil. A type of MRI coil in which orthogonal coils are used. The power deposition is half that of linear coil architecture (Brown et al. (2014)). See Coil Polarization.

Quench. A condition in a superconducting magnet in which resistance develops in the main magnet coil, which heats the coil, which increases the resistance, which causes more heating, etc. This develops a vicious cycle leading to loss of the field and boiling of the cryogens. In the event of a quench breach cryogens may be discharged external to the venting system, which may be hazardous. Quenches can be controlled or uncontrolled. Controlled quenches may be performed by vendor service engineers for various reasons. Uncontrolled quenches may be initiated by pressing the quench button or may be spontaneous, for example due to an earthquake. Uncontrolled quenches have the potential to damage the coil as there is a rapid temperature change inside the magnet on the order of 500 °F on the order of seconds (ACR Committee on MR Safety (2020)).

Although the changing field during a quench is not thought to have adverse health effects, it is considered advisable to evacuate the scanner room prior to or during a quench. It may take on the order of a minute or more for the field ro dissipate in an uncontrolled quench

Quenches may be accompanied by a loud noise. If the quench venting structures work properly the boiled cryogens will be safely vented to the outside. In a quench breach, however, cryogens leak out of the venting system. A fog may appear in the room, and there may be a rapid decrease in room temperature. Importantly, escaping cryogens can displace/dilute the oxygen in the room representing an asphyxiation hazard. See Cryogen Hazards.

After a quench it is necessary to ensure there is no residual field present before introducing potentially unsafe objects into Zone III or IV (ACR Committee on MR Safety (2020)).

Quench Breach. In the event of a quench, a quench breach occurs when the cryogens escape into the space outside the scanner and quench pipe, for example into the scanner room. Explosive breaches may occur if the quench pipe is blocked, and cryogens are unable to escape with associated pressure buildup (ACR Committee on MR Safety (2020)).

Quench Button. A button in the scanner control room which will cause a quench of the magnet. It takes on the order of a minute or more for the field to dissipate. A quench initiated by the quench button is not a controlled quench and could damage the magnet as well as be costly in terms of the cost to replace the cryogens and in terms of downtime. Therefore, the quench button should only be pressed in emergencies such as an individual pinned to the magnet by a metallic object or uncontrolled fire in the scanner room. The scanner room should be evacuated to the extent possible during a quench. After the quench MR personnel should verify that the field has completely dissipated (ACR Committee on MR Safety (2020)).

Quench Pipe. A pipe which safely guides boiled off cryogens from the magnet to the outside of the building in the event of a magnet quench. Note that the quench pipe is Zone IV in the scanner room and then Zone III throughout its remaining course to the outside where it exits the quench vent. The quench pipe should be inspected yearly as blockage could result in a quench breach or explosion (ACR Committee on MR Safety (2020)).

Quench Vent. A vent on the outside of a building which vents cryogens to the outside in the event of a quench. This should be designed so that it is unlikely to be obstructed, for example from precipitation or animals. The quench vent is Zone III and must have appropriate signage and an appropriate safety exclusion zone. The quench exclusion zone should not contain serviceable equipment or windows/doors. Individuals entering this area should be properly trained on the hazards of the area (ACR Committee on MR Safety (2020)).

R

Radiation. In the context of MR safety, radiation refers to a mechanism of patient cooling whereby patients can emit heat from their body, or radiation can refer to electromagnetic rf radiation (Halliday et al. (2014), Zangwill (2013)).

(*MRSE*) Radiation is the emission of energy in the form of motion of electromagnetic waves caused by accelerating charges (Zangwill (2013)).

Radiofrequency (rf). Frequency of electromagnetic radiation ranging from 10's of kHz to 100's of GHz. The frequencies used in MRI are, for example, around 64 MHz for a 1.5 T scanner and 128 MHz at 3 T (Collins (2016)).

RadSite. One of the accrediting organizations designated to accredit MRI sites under MIPPA (Clausen et al. (2018)).

Receive Only Coil. A coil which can only receive the signal from the patient during an MRI examination and cannot be used for excitation. Therefore, receive only coils are used with the built in body transmit coil (Brown et al. (2014)).The safety significance is that much of the central volume of anatomy inside the bore is exposed to rf with the built in body transmit coil, as opposed to a local transmit receive coil, which only excites the local area, e.g. the head.

Receiver Bandwidth. Range of frequencies over which the signal is received and properly characterized (Brown et al. (2014)). The receiver bandwidth is a function of the sampling rate as dictated by the Nyquist Theorem.

Reflection of Electromagnetic Radiation. Reflection of electromagnetic radiation can occur at interfaces (Griffiths (2017)). Reflection of rf radiation can cause burns. See Reflective Burns.

(*MRSE*) Simple example of wave reflection (Halliday et al. (2014))

Given an electric field of an incident electromagnetic wave propagating along z oriented in the x direction (See Electromagnetic Wave)

$$\mathbf{E}_I = E_A \cos(kz - \omega t)\,\widehat{\mathbf{x}} \tag{R.1}$$

If the wave strikes a perfect conductor with the electric field at normal incidence, then the reflected wave will be

$$\mathbf{E}_R = -E_A \cos(kz + \omega t)\,\widehat{\mathbf{x}} \tag{R.2}$$

The two waves will then combine

$$\mathbf{E}_{Total} = \mathbf{E}_I + \mathbf{E}_R = 2E_A \sin(kz)\sin(\omega t)\,\widehat{\mathbf{x}} \tag{R.3}$$

$\mathbf{E}_{Total}$ represents a standing wave with an electric field maximum amplitude of $2E_A$, which is twice that of the incident wave.

Reflective Burns. Burns caused by materials on the patient which can reflect rf back toward the patients. Items such as these include objects with metallic backing such as certain types of blankets (Tsai et al. (2015)).

Refrigerator Magnet. The strength of a refrigerator magnet is in the range of 10-100 G (Collins (2016)).

Relative Permeability (μ_r). A name given to the magnetic susceptibility added to 1. $\mu_r = 1 + \chi_m$ (Zangwill (2013)).

Relative Permittivity (ε_r). A name given to the electric susceptibility added to 1. $\varepsilon_r = 1 + \chi_E$. Also called the dielectric constant (Zangwill (2013)).

Relative PNS of Pulse Sequences. EPI > GRE > SE. EPI based sequences such as DWI with the fastest gradient switching on and off have the most risk of PNS, while those such as SE have the least risk.

Relative SAR of Pulse Sequences. FSEIR > FSE > SE > GRE. Those sequences with the most applied 180 degree pulses such as an FSE sequence with an additional 180 degree inversion pulse (FSEIR) have the greatest risk while those with only a single excitation pulse such as a simple GRE sequence have the least.

Resistance (R). A measure of how much a substance opposes the flow of electrical current (Halliday et al. (2014)).

(*MRSE*) For a length of wire l with cross sectional area A with conductance σ_c, the resistance R is

$$R = \frac{l}{\sigma_c A} \tag{R.4}$$

Resistive Magnet. An Electromagnet (Faulkner (2016), ACR Committee on MR Safety (2020)).

Resistive Heating. See Joule Heating.

Resistivity (ρ_c). (*MRSE*) The inverse of the conductivity σ_c (Halliday et al. (2014)).

$$\rho_c = \frac{1}{\sigma_c} \tag{R.5}$$

Resonant LC Burns. A type of rf burn caused by a worst case scenario of magnetic induction when a circuit has matched capacitance and inductance. While this situation could lead to substantial heating, it is statistically unlikely that a match would occur (Dempsey et al. (2001)).

(*MRSE*) In an RLC circuit with an alternating current, the worst case magnetic induction heating will occur when the circuit with a current of frequency ω is in resonance so that

$$\omega = \frac{1}{\sqrt{LC}} \tag{R.6}$$

where L is the inductance and C is the capacitance.

rf. See Radiofrequency.

rf Coils. Coils which generate the rf field. These can be transmit only, receive only, or transmit/receive.

Receive only coils use the built in body transmit coil for excitation.

(*MRSE*) Coils are labeled according to the IEC labeling convention (International Electrotechnical Commission (2015), McRobbie (2020)):

Figure R.1: Transmit Only Coils.

Figure R.2: Transmit/Receive Coils.

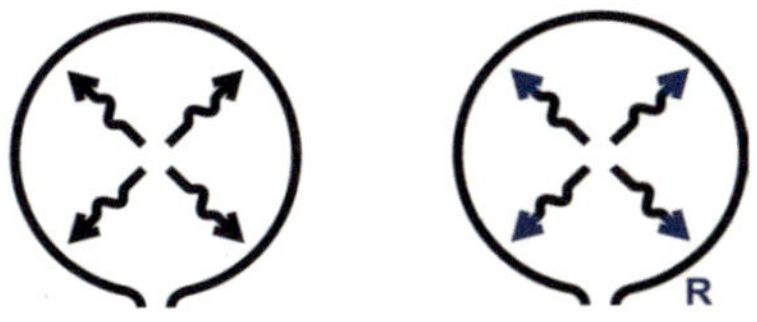

Figure R.3: Receive Only Coils

rf Pulse A short pulse of radiofrequency energy used to excite the spins in the patient as part of obtaining an MR image (Brown et al. (2014)).

rf Heating. Heating due to the applied rf pulses during an MRI acquisition. Heating may be mediated through the magnetic field component or through the electric field component of the electromagnetic field (Dempsey et al. (2001), Dempsey and Condon (2001), Schaefer (2014)).

(*MRSE*) As an illustration of some of the factors involved, the heating of a linear conductor in a linearly polarized birdcage rf coil is calculated using a simple model (Schaefer (2014)). Assuming resonant effects are negligible, the heating of a linear conductor of length l located entirely within the rf transmit coil and oriented with an angle ϕ with respect to the z-axis is considered. The electric field of the applied rf $\mathbf{E}_1$ is along z.

For a birdcage coil the $\mathbf{E}_1$ field is (see E1)

$$\mathbf{E}_1 = E_1\widehat{\mathbf{z}} = \omega y B_1\widehat{\mathbf{z}} \tag{R.7}$$

The induced voltage will be

$$V = \int_0^l \mathbf{E}_1 \cdot d\mathbf{l} = E_1 l \cos\phi \tag{R.8}$$

then

$$V = E_1 l \cos\phi = \omega y B_1 l \cos\phi \tag{R.9}$$

If the conductor contacts the patient through an area A, the current density will be related to the impedance Z of the object to the body as $I = JA$. Through Ohm's Law $V = IZ = JAZ$,

$$J = \frac{V}{AZ} = \frac{\omega y B_1 L \cos\phi}{AZ} \tag{R.10}$$

The associated induced electric field is $E_{ind} = J/\sigma_c$ and the local SAR may be calculated during rf application as (see Specific Absorption Rate)

$$\mathrm{SAR} = \int_V \frac{1}{m}\frac{dP_t}{dV}dV = \frac{\sigma_c \overline{E_{ind}^2} V}{m} = \frac{\sigma_c {E_{ind}}^2 V}{2m} = \frac{\sigma_c {E_{ind}}^2}{2\rho} \tag{R.11}$$

Substituing Eq. R.9 gives

$$\mathrm{SAR} = \frac{\omega^2 y^2 B_1^2 L^2 \cos^2\phi}{2\sigma_c \rho A^2 Z^2} \tag{R.12}$$

The duty cycle f_d of a pulse sequence represents the fraction of time the rf energy is on during a pulse sequence, so that the time averaged SAR is

$$\overline{\mathrm{SAR}} = \frac{\omega^2 y^2 B_1^2 L^2 f_d \cos^2\phi}{2\sigma_c \rho A^2 Z^2} \tag{R.13}$$

To a first approximation

$$\Delta T = \frac{\mathrm{SAR}\Delta t}{C_p} \tag{R.14}$$

where C_p is the heat capacity, T is temperature, and t is time.

Therefore if there is a metallic conductor, in which resonant effects are neglible, the degree of heating increases with the length of the device and decreases with the area of contact with the patient. The degree of heating also increases with frequency and hence field strength as well as duty cycle. In addition, heating will be minimized along the center of the coil at $y = 0$, but increases as y increases toward the periphery of the coil. In addition, because of the $\cos\phi$ dependence, linear objects oriented perpendicular to the bore also have heating at a minimum (Schaefer (2014)).

rf Heating Testing. A standardized test to determine how much an implant heats under test conditions and might heat within the body under a specified set of scan conditions (ASTM F2182-19e2 (2019)). A device with dimensions of less than 2 cm in all directions may not need to be tested since the associated change in temperature is expected to be less than 2 °C over 60 minutes of exposure at 1.5 T and 3 T (ASTM F2182-19e2 (2019)). This condition is not valid if multiple replicas of the device (are implanted within 3 cm of each other.

(*MRSE*) Described in ASTM publication F2182-19e2. The implant is placed in a standard phantom with gelled saline (conductivity 0.47 S/m and dielectric constant 80). The temperature is measured at the implant as a function of time in an area of known background SAR. The local background RF exposure without the implant should be chosen so that it is high enough to heat the implant free region to at least 10 times the precision of the temperature probe used for measurement, for example 1 °C if the precision is 0.1 °C. Pilot experiments can be used to determine areas of placement to measure the maximum rise. The ΔT on or near the implant is measured at several locations during exposure as well as at a reference location at least 30 cm from the implant. The implant is then removed and temperature measurements (with a temperature probe) or electric field measurements (with an electric field probe) are repeated at the same locations used previously with the implant in place. The local background SAR at the reference probe is used to verify that the same RF exposure conditions are applied in the various steps. Local background SAR may be calculated without the implant from the temperature rise using

$$\frac{\Delta T}{\Delta t} = \frac{\text{SAR}}{C_p} \tag{R.15}$$

where C_p is the heat capacity, T is temperature, and t is time using $C_p = 4150\,J/kg$-K or from a direct measurement of the electric field from

$$\text{SAR} = \frac{\sigma_c {E_A}^2}{2\rho} \tag{R.16}$$

where σ_c is the conductivity of the gelled saline, E_A is the magnitude of the peak electric field, and ρ is the density of the gel.

A reference implant may be used to determine local SAR.

Rheobase (I_r). (*MRSE*) The minimum current of infinite duration required to cause nerve depolarization (McRobbie (2014)). See Strength Duration Curve.

Right Handed Coordinate System. (*MRSE*) Coordinate system in which $+x$ is along the index finger, $+y$ is along the right long finger, and $+z$ is along the right thumb when these digits are held orthogonally. For example, with respect to

MRI, for a horizontal bore system the positive z-axis points from table into the bore and the positive y-axis points upward, with the positive x-axis pointing right when the patient is supine head first (Collins (2016), Griffiths (2017)).

Rise Time. The period of time required for a time varying gradient to reach its peak level. Usually on the order of 200 μs (Brown et al. (2014)).

Room Oxygen Monitoring. A monitor can be used to measure the oxygen level in the scan room. During a quench if the oxygen level is less than 19.5% a self contained breathing apparatus should be used to enter the room (Occupational Safety and Health Administration Standard 29 CFR 1915.12). It should be noted that room oxygen monitoring is not specifically recommended at this time by the 2020 ACR on MR Safety (ACR Committee on MR Safety (2020)).

Room Pressure Relief. In the event of a quench breach in which cryogens are discharged into the room potentially trapping those inside from increased pressure against the main door, scanner rooms should be designed with a trapdoor to relieve the pressure (Junk (2014)).

S

SAE. See Specific Absorbed Energy.

Safety Marking. Specified by ASTM publication F2503-20. This includes a discussion of marking of objects as MR Safe, MR Conditional, and MR Unsafe (ASTM F2503-20 (2020)).

(*MRSE*) The marking should not compromise the use of the object and should be of sufficient durability to serve the life of the object. For Conditional objects the conditions for safe use should be listed.

The acceptable labels are:

Acceptable MR Safe labeling options include:

Figure S.1: MR Safe Labels.

Acceptable MR Conditional labeling options include:

Figure S.2: MR Conditional Labels.

Acceptable MR Unsafe labeling options include:

Figure S.3: MR Unsafe Labels.

Items should be tested in accordance with the procedures published by the ASTM or ISO as applicable to the specific device.

Safety of Future Scan Given Prior MRI. A history of a safe MRI scan of an implant or foreign body cannot not necessarily be used to predict the safety of a future scan. Since safety factors such as the field strength, maximum spatial gradient, slew rate, and pulse sequences vary between scanners and examinations, and since the forces on an implant or foreign body can depend upon the orientation of the patient with respect to the magnet or rate of motion of the patient with respect to the magnet, it cannot therefore be assumed that a history of prior uneventful MRI will guarantee the safety of an MRI in the future (ACR Committee on MR Safety (2020)).

Safety Labeling. See Safety Marking.

SAR. See Specific Absorption Rate.

Saturation. See Magnetic Saturation.

Saturation Recovery. Pulse sequence consisting of multiple 90 degree pulses (Brown et al. (2014)).

Scalar. A quantity represented by a number. This is distinguished from a vector characterized by a number and a direction. Energy is a scalar, while force is a vector. The magnitude of a vector is a scalar (Griffiths (2017)).

Scalar Field. (*MRSE*) A collection of points where at every point in space there is a property associated with a number (Griffiths (2017)).

Scanner Orientation. (*MRSE*) The orientation of a horizontal field scanner with respect to field and coordinate system varies by manufacturer, although the orientation used herein is that corresponding to (Collins (2016)) with the $+z$-axis extending into the bore from the table. Note that some scanners have the $+z$-axis extending from the bore toward the end of the table and the $+x$-axis directed toward the left with the y-axis still extending toward the ceiling (Brown et al. (2014), ASTM F2052-15 (2015))

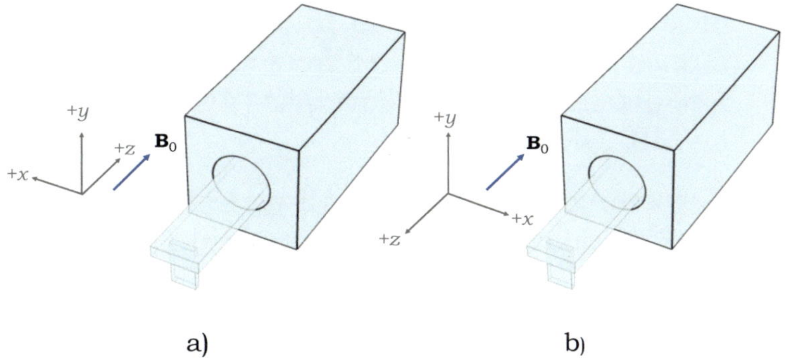

Figure S.4: Scanner orientation with respect to field and coordinate system. a) (Collins (2016)). b) (Brown et al. (2014), ASTM F2052-15 (2015)).

Screening. In MR safety, a process whereby patients, staff, and other individuals are evaluated regarding their safe entrance into the MR environment and any subsequent scanning. Screening usually involves written questions and interactive verbal questions, but may also involve physical examination, ferromagnetic detector screening, and radiological screening (Saraswat and Smith (2014), ACR Committee on MR Safety (2020)).

Screening of Family and Other Non-MR Personnel. Screening of any family or other Non-MR personnel should be identical to that for patients. The 2020 ACR Manual on MR Safety has a section indicating that if orbital screening is necessary in any Non-MR personnel accompanying the patient, Level 2 personnel should counsel the individual regarding the need for orbital radiographs. If the individual wishes to proceed past the 5 G line, and the Level 2 physician thinks it is medically indicated for the individual to proceed, written informed consent is required from the individual before obtaining orbit radiographs (ACR Committee on MR Safety (2020)).

Screening of Patients. All patients undergoing MRI examinations must be screened (ACR Committee on MR Safety (2020)). Patients or their appropriate representative must fill out a written questionnaire. Forms such as those provided by the American College of Radiology (ACR) or Institute for Magnetic Resonance Safety and Research (IMRSER) are commonly used, or the forms used are typically based on those from the ACR or IMRSER (Saraswat and Smith (2014)). These have checkboxes regarding prior surgery, implants, and foreign bodies. All boxes must be checked; no blank answers should be accepted. The questionnaire also inquires about the possibility of pregnancy, any prior examinations, or problems with any prior MRI or contrast. The patient or representative must then sign the form. The form is then reviewed with the patient or representative by Level 2 personnel who interactively does a verbal screening. The screening form must be signed by the Level 2 individual as well as the patient or their representative. Conventional metal detectors are not recommended as an adjunct for screening; however, ferromagnetic detectors are recommended by the ACR as an adjunct to screening. Imaging such as radiography or CT may also be appropriate as an adjunct measure in screening. If there is definitive history of an implant or foreign body, effort should be made to obtain sufficient information to justify a decision to proceed with or deny scanning. Inpatients may be screened in the patient room for hospital-based MRI facilities, although all screenings are to be double-checked and verified by Level 2 MR Personnel prior to scanning. Unresponsive or unreliable patients follow a different procedure (ACR Committee on MR Safety (2020)). See Screening of Unconscious Patients.

Screening of Personnel. Initial screening of personnel as part of their employment should be identical to screening of patients, although if any MR personnel has any injury or procedure in which a new ferromagnetic object or new device is introduced into them, this must be immediately reported to the MR medical director. Repeat screening and assessment of safety of the MR personnel in the

MR environment should then be performed (ACR Committee on MR Safety (2020)). See Screening of Patients.

Screening of Unconscious Patients. MRI may be necessary for patients who are unconscious or unreliable and cannot provide their own histories. An attempt should then be made to obtain such history from a reliable surrogate, such as a family member, regarding their history of their metal exposure. If the clinical indication is such that it is not reasonable to wait until such history could be obtained, it is recommended that patients have plain films of the head, neck, chest, abdomen, pelvis, as well as upper arms and thighs if prior imaging of these areas is not already available in a reasonably recent time frame. If there are obvious post-traumatic changes to the distal extremities, those areas should also undergo radiography prior to MRI (ACR Committee on MR Safety (2020)).

Second Level Controlled Operating Mode. Mode of MRI scanner operation in which the output may cause significant risk. Use of this mode requires institutional approval. These levels are above the IEC limits for First Level Controlled Operating Mode (Keevil (2014), Schaefer (2014)). See Exposure Limits.

SED. See Specific Energy Dose.

SEL. See Specific Energy Limit.

Sedation. A continuum of sensory and functional depression including a continuum of levels categorized as minimal sedation, moderate sedation, deep sedation, and general anesthesia (Reinking (2014)).

Some patients require sedation to varying degrees for various reasons when undergoing an MRI scan.

Shape Factor. A numeric factor used to weight a pulse in terms of its SAR contribution based on waveform (Schaefer (2014)).

Sharp Points. Areas of implant, foreign body, or anatomic sharp edges can be problematic from the standpoint of MRI safety since at these points there is increased charge density and electric field which can lead to tissue injury (Zangwill (2013)).

(*MRSE*) Simple model: Suppose there are two hollow conducting spheres 1 and 2 separated by a sufficient distance that the charge on one does not significantly affect the other. The two spheres are connected by a thin conducting wire. The charge is confined to the surface of the conductor with an equipotential along the system.

The electric potential can be considered separately for each sphere. Calculating the electric potentials V_1 and V_2 gives

$$V_1 = \frac{q_1}{4\pi\varepsilon_0 r_1} = V_2 = \frac{q_2}{4\pi\varepsilon_0 r_2} \tag{S.1}$$

where q_1 and r_1 are the charge and radius of sphere 1 and q_2 and r_2 are the charge and radius of sphere 2, and ε_0 is the permittivity of free space. From this

$$q_2 = \frac{q_1 r_2}{r_1} \tag{S.2}$$

Note that if $r_2 > r_1$, then $q_2 > q_1$, and the to-

Figure S.5: Model system with two hollow conducting spheres with thin conducting wire between them. Note that the total charge on the larger sphere is bigger than that of the smaller sphere, but that the charges on the smaller sphere are closer together, that is, the charge density on the smaller sphere is greater than that for the larger sphere.

tal charge on the larger sphere is greater, but the charge density for a sphere $j = 1$ or $j = 2$ is

$$\sigma_{q_j} = \frac{q_j}{A_j} = \frac{q_j}{4\pi r_j^2} \tag{S.3}$$

where $A_j = 4\pi r_j^2$ is the suface area of sphere j, so that in this case for $r_2 > r_1$

$$\sigma_{q_1} = \frac{\sigma_{q_2} r_2}{r_1} \tag{S.4}$$

and $\sigma_{q_1} > \sigma_{q_2}$, so that the charge density of the smaller sphere is greater. To then compute the electric fields, $\mathbf{E}_1$ of sphere 1 and $\mathbf{E}_2$ of sphere 2, Gauss's Law gives

$$\frac{Q_{e_j}}{\varepsilon_0} = \frac{\sigma_{q_j} A_j}{\varepsilon_0} = \oint_S \mathbf{E} \cdot d\mathbf{A} = E_j A_j \tag{S.5}$$

so that

$$\frac{\sigma_{q_1}}{\sigma_{q_2}} = \frac{E_1}{E_2} \tag{S.6}$$

This indicates that electric field is therefore also higher at regions with higher curvature. This is depicted in Figure S.5 where the smaller sphere with a higher radius of curvature has a higher surface charge density. The electric field may then generate a current density which can lead to a current and resistive heating with associated tissue damage.

SI Units. International System of Units. SI is an abbreviation of the French Systeme International, and is a system of measure with seven base units of Ampere, seconds, meters, Kelvin, candela, kilogram, and mole. Other derived units relevant to MRI are Tesla, Hertz, Newton, Joule, Watt, Coulomb, Volt, Henry, degree Celsius, Farad, and Ohm (Halliday et al. (2014)).

Signal-to-Noise (SNR). Numerical ratio of signal intensity to noise intensity. The signal-to-noise is given by (Brown et al. (2014), McRobbie et al. (2017))

$$\frac{SNR}{voxel} \propto \frac{\Delta x \Delta y \Delta z \sqrt{N_x}\sqrt{N_y}\sqrt{N_z}\sqrt{N_{acq}}}{\sqrt{BW_{read}}} \tag{S.7}$$

where Δx, Δy, and Δz are the slice thicknesses in the three dimensions x, y, and z, N_x, N_y, and N_z are the number of points in each dimension, N_{acq} is the number of acquisitions, and BW_{read} is the receiver bandwidth.

Slew Rate. The rate at which the time varying gradients turn on. They ramp up from zero to the gradient strength demanded for the FOV. This means there will be a magnetic field changing rapidly in time, which can induce currents by Faraday's Law within patients or medical implants. This may cause PNS or device malfunction. A typical maximum slew rate might be 200 T$/m/s$ (Ansorge and Graves (2016)).

(*MRSE*) The slew rate SR is given by (Brown et al. (2014), McRobbie (2014), McRobbie et al. (2017))

$$SR = \frac{(G_+ + G_-)}{t_{\min}} = \frac{G_{Max}}{t_r} \tag{S.8}$$

where G_+ and G_- are the maximum and minimum absolute value amplitudes of the gradient and t_{min} is the minimum time to transition between them. G_{Max} is the maximum gradient amplitude and t_r is the rise time.

Example: For a maximum gradient amplitude of 40 mT$/m$ with a minimum rise time of 200 μs, calculate the slew rate.

$$
\begin{aligned}
SR &= \left(\frac{40\,\mathrm{mT}}{m}\right)\left(\frac{1}{200\,\mu s}\right)\left(\frac{1\,\mathrm{T}}{1000\,\mathrm{mT}}\right)\left(\frac{1\times 10^{6}\,\mu s}{1\,s}\right) \\
&= 200\,\mathrm{T}/m/s
\end{aligned} \tag{S.9}
$$

Slice Thickness. In terms of MRI safety, slice thickness is one of the things which determines gradient strength (and hence dB/dt) and also SAR (through the transmitter bandwidth) (Brown et al. (2014)).

$$\Delta z = \frac{\Delta\omega_{rf}}{\gamma G_{ss}} \tag{S.10}$$

where $\Delta z = FOV_{ss}$ is the slice thickness, γG_{ss} is the slice select gradient, and $\Delta\omega_{rf}$ is the transmitter bandwidth.

Solenoid. A circular or coiled loop of wire. Magnetic fields are generated when current is passed through the solenoid (Halliday et al. (2014)).

Spatial Gradient. Rate at which the main magnetic field changes with distance. Along the z-axis of the magnetic field along the axis of the bore close to the magnet, this is approximated as dB_0/dz. The units are T/m or G/cm. 1 T/m = 100 G/cm (Shellock et al. (2011), ASTM F2052-15 (2015)). The force on an unsaturated magnetic object depends upon the force product $B_0 dB_0/dz$ and on $m_S dB_0/dz$ for a saturated object, where m_S is the saturated magnetic moment.

(*MRSE*) The more precise meaning is the magnitude of the gradient of the magnitude of the main magnetic field

$$|\nabla |\mathbf{B}_0|| = |\nabla B_0| \tag{S.11}$$

which for the situation described of being along the axis bore near the maximum value of the spatial gradient becomes dB_{0_z}/dz. This is the situation for displacement force testing. See Magnetic Displacement Force and Displacement Force Testing.

Specific Absorbed Energy (SAE). See Specific Energy Dose.

Specific Absorption Rate (SAR). Rate at which rf energy is deposited into tissue (Brown et al. (2014) Schaefer (2014), Ansorge and Graves (2016)). Usually given in terms of W/kg. The determination of SAR requires calculation from an assumed model and therefore potentially varies in accuracy. SAR calculations are also vendor specific and proprietary. The estimated SAR is displayed on the scanner for a given pulse sequence calculated from the patient's body weight input into the scanner. It is helpful to remember the following general proportionalities:

$$SAR \propto B_0^2 \propto B_1^2 \propto \omega_0^2 \propto \omega_1^2 \propto \alpha_p^2 \propto \frac{1}{T_R} \tag{S.12}$$

for specific absorption rate (SAR), magnitude of the static field (B_0), magnitude of applied rf field (B_1), Larmor frequency (ω_0), applied rf frequency (ω_1), pulse flip angle (α_p), and repetition time (T_R).

Example: All other things remaining equal, how does the SAR change in going from 1.5 T to 3 T? Since SAR is proportional to the square of the field, the SAR would quadruple.

(*MRSE*) It is worth having some understanding of how SAR is calculated based on a simple model, although the exact SAR calculation used by vendors is proprietary. Following the example given in Brown et al. (2014), Schaefer (2014), and Ansorge and Graves (2016) a simple model of the patient's body as a sphere can be constructed. For a sphere of radius R_s, the radius of any ring on the sphere less than R_s can be obtained in spherical coordinates as $r' = r\sin\theta$. By Faraday's Law assuming a linearly applied field oriented perpendicular to the plane of the ring

$$E = \frac{1}{l}\oint_C \mathbf{E}\cdot d\mathbf{l} = -\frac{1}{l}\frac{\partial}{\partial t}\int_S \mathbf{B}_1 \cdot d\mathbf{A} \qquad \text{(S.13)}$$

where l is the circumference around the circle r'. Assuming an applied field magnitude of $2B_1$ then (Brown et al. (2014), Ansorge and Graves (2016))

$$\begin{aligned}-\frac{\partial}{\partial t}\int_S \mathbf{B}_1 \cdot d\mathbf{A} &= -\frac{\partial}{\partial t}\int_S 2B_1\left(\widehat{\mathbf{x}} cos\omega_0 t\right) d\mathbf{A} \\ &= -A\left|\frac{\partial \mathbf{B}_1}{\partial t}\right| = A\omega_0 B_1 sin\omega_0 t\end{aligned} \qquad \text{(S.14)}$$

Using $A = \pi r^2 \sin^2\theta$ and $l = 2\pi r sin\theta$ the magnitude of the electric field can be calculated as

$$
\begin{aligned}
E &= \frac{A\omega_0 B_1 sin\omega_0 t}{l} = \omega_0 B_1 r \sin\theta sin\omega_0 t \\
&= E_A sin\omega_0 t
\end{aligned}
\tag{S.15}
$$

The power deposited is expressed by the mean squared time averaged ohmic power density dP_t/dV (Brown et al. (2014) Ansorge and Graves (2016)) which from Eq. S.15 is (See Power Density)

$$
\frac{dP_t}{dV} = \overline{\sigma_c E^2} = \frac{1}{2}\sigma_c {E_A}^2 = \frac{1}{2}\sigma_c \omega_0^2 B_1^2 r^2 \sin^2\theta \tag{S.16}
$$

On integration of Eq. S.16 over the volume of the sphere the time averaged ohmic power P_t is

$$
\begin{aligned}
P_t &= \int_V \frac{1}{2}\sigma_c {E_A}^2 dV \\
&= \int_0^{\pi}\int_0^{2\pi}\int_0^{R_s} \frac{1}{2}\sigma_c {E_A}^2 r^2 \sin\theta dr d\phi d\theta \\
&= \int_0^{\pi}\int_0^{2\pi}\int_0^{R_s} \frac{1}{2}\sigma_c \omega_0^2 B_1^2 r^4 \sin^3\theta dr d\phi d\theta \\
&= \frac{4}{15}\pi\sigma_c\omega_0^2 B_1^2 R_s^5
\end{aligned}
\tag{S.17}
$$

For a given rf pulse using $\alpha_p = \gamma B_1 \tau_p$

$$
P_t = \frac{4\pi\sigma_c\omega_0^2\alpha_p^2 R_s^5}{15\gamma^2\tau_p^2} \tag{S.18}
$$

The total energy deposited over the time of the pulse will be

$$\mathcal{E} = P_t \tau_p = \frac{4\pi \sigma_c \omega_0^2 \alpha_p^2 R_s^5}{15\gamma^2 \tau_p} \tag{S.19}$$

Given power P_t deposited in a given mass m of density ρ and volume $V = 4\pi R_s^3/3$

$$\text{SAR} = \frac{P_t}{m} = \frac{P_t}{\rho\left(\frac{4}{3}\pi R_s^3\right)} = \frac{B_1^2 \sigma_c \omega_0^2 R_s^2}{5\rho} = \frac{B_1^2 \sigma_c \gamma^2 B_0^2 R_s^2}{5\rho} \tag{S.20}$$

using $\omega_0 = \gamma B_0$ (Brown et al. (2014)). In addition, again using $\alpha_p = \gamma B_1 \tau_p$

$$\text{SAR} = \frac{\sigma_c \omega_0^2 \alpha_p^2 R_s^2}{5\rho\gamma^2\tau_p^2} = \frac{\sigma_c B_0^2 \alpha_p^2 R_s^2}{5\rho\tau_p^2} \tag{S.21}$$

The time averaged power, $\overline{P}$, represents the total energy $\mathcal{E}$ over the total time, so that when the time is taken over the repetition time T_R (Brown et al. (2014))

$$\overline{P} = \frac{\mathcal{E}}{T_R} = P_t \frac{\tau_p}{T_R} = \frac{4\pi\sigma_c\omega_0^2\alpha_p^2 R_s^5}{15\gamma^2\tau_p T_R} = \frac{4\pi\sigma_c B_0^2 \alpha_p^2 R_s^5}{15\tau_p T_R} = \frac{4\pi\sigma_c B_0^2 \gamma^2 \tau_p R_s^5}{15 T_R} \tag{S.22}$$

and therefore for the time averaged SAR

$$\overline{\text{SAR}} = \frac{\sigma_c \omega_0^2 \alpha_p^2 R_s^2}{5\rho\gamma^2\tau_p T_R} = \frac{\sigma_c B_0^2 \alpha_p^2 R_s^2}{5\rho\tau_p T_R} = \frac{\sigma_c B_0^2 \gamma^2 B_1^2 \tau_p R_s^2}{5\rho T_R} \tag{S.23}$$

Note several things. First, the SAR depends upon R_s^2 so that larger patients tend to heat more. Second SAR depends on the square of both B_0 and the flip angle α_p so that doubling the field quadruples the SAR. Going from a 90-degree pulse to a 180-degree pulse also quadruples the SAR. The SAR depends on α_p^2/τ_p^2 and hence B_1^2. The $\overline{\text{SAR}}$ is directly proportional to α_p^2/τ_p as well as $B_1^2\tau_p$ and inversely proportional to T_R.

Example: How does the $\overline{\text{SAR}}$ for a 90-degree pulse with a pulse time of 6 ms and T_R of 2000 ms compare to that of a 20-degree pulse with a pulse time 2 ms and T_R of 300 ms both done at 1.5 T on the same patient?

$$\begin{aligned} \frac{\overline{\text{SAR}_2}}{\overline{\text{SAR}_1}} &= \left(\alpha_p^2/(\tau_p T_R)\right)_2 \Big/ \left(\alpha_p^2/(\tau_p T_R)\right)_1 \\ &= 90^2/(6 \cdot 2000) \Big/ 20^2/(2 \cdot 300) \approx 1 \end{aligned} \tag{S.24}$$

Despite significant differences in acquisition parameters, the SAR is roughly the same.

In the typical pulse sequence, there is a series of pulses and delays in a pulse train as shown in Figure S.6.

It is also instructive to consider in conjunction with Figure S.6 the form (Collins and Smith (2001))

$$\overline{\text{SAR}} = \frac{\sum_{i=1}^{n} \mathcal{E}_i}{mT_T} = \frac{\sum_{i=1}^{n} \text{SAR}_i \tau_{p_i}}{\sum_{i=1}^{n} \tau_i + t_{p_i}} \tag{S.25}$$

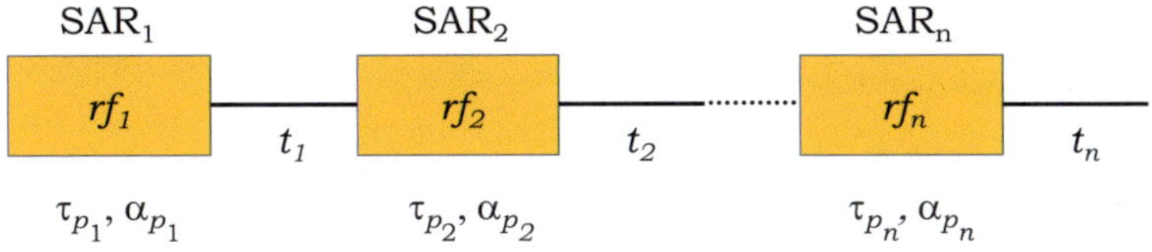

Figure S.6: Pulse train. Individual rf pulses are shown with individual rf pulses with individual SAR$_i$ with flip angles α_{p_i} pulse times τ_{p_i} with intervening times t_i. The t_i are times such as half echo times $TE/2$, inversion times T_{IR}, etc.

where the $\mathcal{E}_i$ represent the energy deposited by each rf pulse, SAR$_i$ the individual SAR of each pulse, and τ_{p_i} the pulse time of each pulse. T_T is the total time under consideration, which is a sum of the pulse times as well as the time periods t_i during which no rf is applied.

The effects of various parameters on SAR can be examined using Eq. S.25. If T_T is equated to T_R, then increasing T_R will decrease $\overline{\text{SAR}}$. This would include the effect of increasing the echo time TE by virtue of increasing the t_i. Considering the effect of multiple acquisitions $\overline{\text{SAR}}$ will be unchanged, as for N acquisitions

$$\overline{\text{SAR}} = \frac{\sum_{i=1}^{n} N\text{SAR}_i \tau_{p_i}}{NT_R} = \frac{\sum_{i=1}^{n} \text{SAR}_i \tau_{p_i}}{T_R} \tag{S.26}$$

At least to a first approximation, NEX does not alter SAR, although the total energy deposited does increase with NEX. Note that the more rf pulses which occur during a given T_R the higher the calculated SAR. Therefore SAR increases, for example, with echo train length. In particular, for number of echoes N_E and number of slices N_S (McRobbie (2020))

$$SAR \propto \frac{B_0^2 \alpha_p^2 N_E N_S}{\tau_p T_R} \tag{S.27}$$

Specific Energy Dose (SED). Also known as specific absorbed energy (SAE). Units of J/kg (Schaefer (2014),/, ACR Committee on MR Safety (2020)). A measure of the total energy dose to the patient. Some manufacturers have implemented SED limits on their MR scanners, for example 14,440 J/kg (4 W/kg for one hour), to limit patient core temperature elevations, discomfort, or physiological stress related to thermal deposition during prolonged imaging. Some manufacturers have incorporated shutdowns of the scanner when a particular SED is reached, with some manufacturers also incorporating a lockout period such as 24 hours before allowing continued scanning. It should be possible, however, to still safely scan patients using sufficiently long cool down periods between sequences even for a high total SED (ACR Committee on MR Safety (2020)).

Speed of Light (c). The speed of light in vacuum is $c = 1/\sqrt{\mu_0 \varepsilon_0} = \nu_0 \lambda_0 = 3 \times 10^8 \; m/s$ (Halliday et al. (2014)).

Spherical Coordinates. (*MRSE*) Coordinate system based on r, θ, and ϕ base vectors. The relationship of Cartesian to spherical coordinates is given by (Griffiths (2017))

$$\widehat{\boldsymbol{r}} = sin\theta cos\phi \widehat{\mathbf{x}} + sin\theta sin\phi \widehat{\mathbf{y}} + cos\theta \widehat{\mathbf{z}} \tag{S.28}$$

$$\widehat{\boldsymbol{\theta}} = cos\theta cos\phi \widehat{\mathbf{x}} + cos\theta sin\phi \widehat{\mathbf{y}} - sin\theta \widehat{\mathbf{z}} \tag{S.29}$$

$$\widehat{\boldsymbol{\phi}} = -sin\phi \widehat{\mathbf{x}} + cos\phi \widehat{\mathbf{y}} \tag{S.30}$$

and

$$\widehat{\mathbf{x}} = sin\theta cos\phi \widehat{\boldsymbol{r}} + cos\theta cos\phi \widehat{\boldsymbol{\theta}} - sin\phi \widehat{\boldsymbol{\phi}} \tag{S.31}$$

$$\widehat{\mathbf{y}} = sin\theta sin\phi \widehat{\boldsymbol{r}} + cos\theta sin\phi \widehat{\boldsymbol{\theta}} + cos\phi \widehat{\boldsymbol{\phi}} \tag{S.32}$$

$$\widehat{\mathbf{z}} = cos\theta \widehat{\boldsymbol{r}} - sin\theta \widehat{\boldsymbol{\theta}} \tag{S.33}$$

In Cartesian coordinates the unit vectors are given by $\widehat{\mathbf{x}}$, $\widehat{\mathbf{y}}$, and $\widehat{\mathbf{z}}$. In spherical coordinates, the unit vectors are denoted $\widehat{\boldsymbol{r}}$, $\widehat{\boldsymbol{\theta}}$, and $\widehat{\boldsymbol{\phi}}$.

Spherical Wave. (*MRSE*) Waves concentrically emanating from a source. Electromagnetic waves are appropriately modeled by spherical waves near a source such as an antenna. At a greater distance the local surfaces may be modeled as plane waves (Rao (2009)).

Spin Echo (SE). An MRI echo in which a 180-degree pulse is sandwiched between a pair of gradients of the same sign and strength. Spin echoes are typically higher SAR than gradient echoes (Brown et al. (2014)).

Staffing of MRI Facilities. There should be a minimum of two MR technologists or one MR technologist and one other MR Personnel (Level 1 or Level 2) in the immediate Zone II through Zone IV MR environment whenever patients are present. At all times the two MR Personnel must be within earshot of each other. In addition, the 2020 ACR MR Safety Manual supports the 2018 Veterans Health Administration Responsibilities Directive 1105.05 for the Medical Facility Director.

This directive ensures that when routinely scheduled patients or research subjects are present in Zones II through IV, there will be a minimum number of MR personnel in Zones III through IV present for safe operation and access control. In particular, when there is one MR machine per Zone III/IV, a minimum of two MR personnel in Zones III through IV are required, with at least one of these personnel designated as Level 2 MR personnel. A temporary exception to this is when MR personnel are interviewing the patient or volunteer or getting them from the waiting or changing areas.

Facilities with two or more MR machines sharing a single Zone III area with at least two machines in use at the same time will have a minimum of one Level 2 MR Personnel for each machine and a minimum of one additional MR personnel (Level 1 or Level 2), so that two machines during scheduled hours would necessitate two Level 2 MR personnel and an additional Level 1 or 2 MR personnel. When only one of multiple machines is in use, there will be a minimum of two MR personnel in Zones III through IV, at least one of which is Level 2 MR Personnel.

Static Field Gradient. See Spatial Gradient.

Stainless Steel. Alloy of iron and chromium. Different types of stainless steel have different magnetic properties and hence interact differently with the MRI scanner's static magnetic field. Austenitic stainless steel typically does not exhibit significant magnetism, whereas ferritic and martensitic steels are magnetic. Ferritic stainless steel has iron and chromium, whereas martensitic has added carbon. Austenitic has added nickel (Dewangan et al. (2015), Lo et al. (2009), Baddoo (2008), Suprihanto (2017)), Cullity and Graham (2011)).

Standing Wave. A wave which oscillates in time but does not travel in space (Rao (2009)).

(*MRSE*) A standing wave $\mathbf{A}(z,t)$ has a form such as

$$\mathbf{A}(z,t) = 2A_0 \sin(kz) \sin(\omega t)\, \widehat{\mathbf{x}} \tag{S.34}$$

where A_0 is the wave amplitude, $k = 2\pi/\lambda$ is the wavenumber, $\omega = 2\pi\nu$ is the angular frequency, t is time, z refers to the z direction, and $\widehat{\mathbf{x}}$ is the unit vector in the x direction. Eq. S.34 describes a wave which oscillates in time, but the amplitude at any point does not travel in space.

A standing wave can be formed by two traveling waves propagating along z oriented in the x direction but moving in opposite directions, for example

$$\mathbf{A}_R = A_0 \cos(\omega t - kz)\, \widehat{\mathbf{x}} \tag{S.35}$$

and

$$\mathbf{A}_L = -A_0 \cos(\omega t + kz)\, \widehat{\mathbf{x}} \tag{S.36}$$

The two waves will then combine

$$\mathbf{A}_{Total_S} = \mathbf{A}_R + \mathbf{A}_L = 2A_0 \sin(kz) \sin(\omega t)\, \widehat{\mathbf{x}} \tag{S.37}$$

Note that the amplitude of the standing wave is twice the amplitude of the traveling waves.

Another important form can be obtained from

$$\mathbf{A}_R = A_0 \cos(\omega t - kz)\, \widehat{\mathbf{x}} \tag{S.38}$$

and

$$\mathbf{A}_L = A_0 \cos\left(\omega t + kz\right) \widehat{\mathbf{x}} \tag{S.39}$$

so that

$$\mathbf{A}_{Total_C} = \mathbf{A}_R + \mathbf{A}_L = 2A_0 \cos\left(kz\right) \cos\left(\omega t\right) \widehat{\mathbf{x}} \tag{S.40}$$

The right and left wave sums may be written in phasor form as

$$\widetilde{A}_{Total_S} = A_0 \left(e^{-ikz} - e^{ikz}\right) = 2iA_0 sin(kz) \tag{S.41}$$

and

$$\widetilde{A}_{Total_C} = A_0 \left(e^{-ikz} + e^{ikz}\right) = 2A_0 cos(kz) \tag{S.42}$$

Note that since $k = 2\pi/\lambda$, Eq. S.41 will have nodes for $z = n\lambda/2$ for $n = 0, 1, 2, 3...$, while Eq. S.42 will have maximum total absolute amplitude antinodes for $z = n\lambda/2$ for $n = 0, 1, 2, 3...$ along z.

Static Magnetic Field. The static magnetic field is the field generated by the currents flowing through the main coils of the magnet, and is usually symbolized $\mathbf{B}_0$. It is called static because it does not change with time although it does change with distance (Brown et al. (2014)). See $\mathbf{B}_0$.

Strategies to Reduce SAR. To reduce the calculated SAR increase TE, increase T_R, decrease echo train length (ETL), decrease flip angle, decrease field strength, increase pulse duration, reduce slice number for a given T_R, increase concatenation, avoid inversion recovery, avoid saturation bands, use gradient echo sequences over SE and FSE, use hyperechoes, and avoid magnetization transfer (Shellock (2020), McRobbie (2020)).

Strategies to Reduce Patient Heating. Strategies to decrease patient heating include use of GRE, SSGRE, or EPI as opposed to SE or FSE, avoid fat saturation pulses, avoid magnetization transfer, avoid inversion recovery, decrease flip angle, increase T_R, increase TE, decrease slices in a given T_R, increase concatenations, increase pulse duration, control room temperature and humidity, interleave high and low SAR acquisitions, use bore ventilation fan, and avoid heavy blankets (Allison and Yanasak (2015), Shellock (2020), McRobbie (2020)).

Strategies to Reduce Peripheral Nerve Stimulation. Avoid echo planar sequences, avoid tissue loops, and use Normal Operating Mode (McRobbie (2020)).

Strength Duration (SD) Curve. Plot of current threshold required to cause nerve stimulation vs. duration of stimulation (McRobbie (2014), McRobbie (2020)). This is relevant to peripheral nerve stimulation which can occur during an MRI scan.

(*MRSE*) Two forms of the strength duration (SD) curve are considered. The first is the Weiss form.

$$I_{th}(\tau) = I_r \left(1 + \frac{C_h}{\tau}\right) \qquad \text{(S.43)}$$

where I_{th} is the threshold generator current, τ is the stimulus duration, and the current $I_r = I_{th}(\infty)$ is known as the rheobase. The constant C_h is called the chronaxie. For $C_h = \tau$, $I_{th} = 2I_r$, so that the chronaxie is the stimulus duration required to double the rheobase as shown in Figure S.7 (McRobbie (2014)).

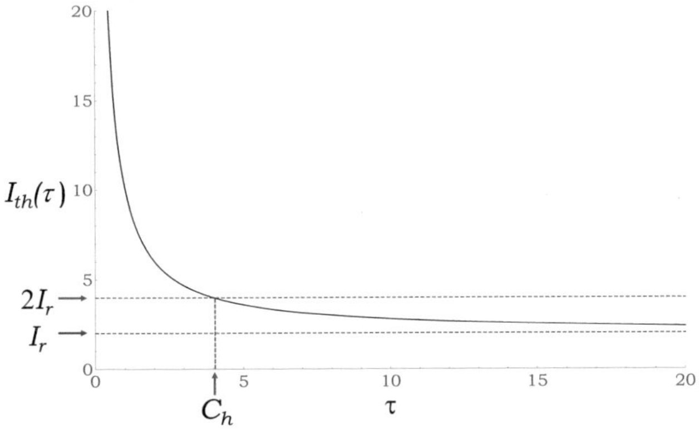

Figure S.7: Strength duration curve (current threshold (I_{th}) vs. stimulus duration (τ)) showing the rheobase (I_r) and chronaxie (C_h). Note that the chronaxie is the stimulus time required to double the rheobase.

Another common form of the SD curve is the Lapicque form (McRobbie (2014)).

$$I_{th}(\tau) = \frac{I_r}{(1 - e^{-\tau/t_c})} \qquad \text{(S.44)}$$

where in this case t_c is a constant called the tissue time constant.

Since for MRI the stimulus arises from dB/dt, the stimulus current can be considered proportional to dB/dt, so that (McRobbie (2012))

$$\left(\frac{dB}{dt}\right)_{th}[\tau] = \left(\frac{dB}{dt}\right)_{r}\left(1 + \frac{C_h}{\tau}\right) \qquad \text{(S.45)}$$

and (McRobbie (2014))

$$\left(\frac{dB}{dt}\right)_{th}[\tau] = \frac{\left(\frac{dB}{dt}\right)_{r}}{(1 - e^{-\tau/t_c})} \qquad \text{(S.46)}$$

Alternatively, one can consider the change ΔB as the stimulus such that (McRobbie (2014))

$$\Delta B(\tau) = \Delta B_{min}\left(1 + \frac{\tau}{C_h}\right) \qquad \text{(S.47)}$$

where $\Delta B_{min} = \Delta B(0)$ is the minimum change in magnetic field for an infinitely short stimulus.

Superconductivity. When a substance is cooled below its critical temperature its resistance drops to zero, and a current, and hence a magnetic field, can persist without a persistent power supply (Coey (2010), Zangwill (2013), Faulkner (2016), McRobbie (2020)).

Superconducting Magnet. Most MRI scanners use a coil immersed in liquid helium. A power supply is connected to the coil and current ramped up to get a given magnetic field, then the power supply is disconnected but the current and magnetic field persist (Brown et al. (2014), Faulkner (2016), McRobbie (2020)). Some older magnets also use liquid nitrogen to surround the liquid helium. Some modern systems are zero boil off devices where any helium which boils off and is recondensed, and do not require helium replenishment(McRobbie (2020)).

Susceptibility. See Electric Susceptibility and Magnetic Susceptibility.

Susceptibility Artifact Distortion in an image related to differences in magnetic susceptibility of two media in the same area of space (Faulkner (2016)). Relevant to MRI safety, implants typically have different susceptibilities from adjacent tissue causing artifact. This is measured in artifact testing as part of FDA labeling (ASTM F2119-07 (2013)). See Artifact Testing. The degree of susceptibility artifact on a magnetic resonance image may also be used to estimate the degree of ferromagnetism of an implant (ACR Committee on MR Safety (2020)).

T

T wave. Final wave of the EKG complex relating to the repolarization of the ventricles. T wave changes occur on the EKG in a strong magnetic field due to flow potentials.

Tattoo. Ornamental drawing on the skin. If the tattoo has been recently placed, there is a chance of bleeding of the image due to magnetic forces from the field. Some tatoos will also heat up. Some colors may be more likely than others to heat. If the tattoo has been placed within 48 hours, there is the potential for smearing/smudging of the tattoo design. The incidence of MRI related problems with tattoos has been reported at 1.5%. A heat sink can be applied to minimize heating (Shellock (2020), ACR Committee on MR Safety (2020)).

Tensor. (*MRSE*) A generalization of a vector which has different magnitudes in different directions. A tensor is used to describe quantities which are anisotropic. Second rank tensors used in MRI can be written as 3 x 3 matrices and are used to describe properties such as magnetic susceptibility when the interactions are different in different directions ((Arfken and Weber (2005), Mueller (2011), Griffiths (2017)).

Tesla (T). SI unit of magnetic flux density, but commonly referred to as field strength. Common MRI strengths are 1.5 T, 3 T, and 7 T. 1 T = 10,000 G. By comparison the Earth's magnetic field is 0.5 G (Halliday et al. (2014), Collins (2016)).

Time Varying Gradients. Gradients varying in time and space applied to form an MR image. These are characterized by a maximum gradient strength as well as a slew rate, which is how fast the gradient turns on and off (Brown et al. (2014)).

TJC. The Joint Commission. Formerly known as the Joint Commission on Accreditation of Healthcare Organizations (JCAHO). A nonprofit organization which accredits and certifies healthcare entities (Clausen et al. (2018)).

Torque. A force which tries to rotate an object (Halliday et al. (2014)). Gravitational torque tries to rotate an object in the Earth's gravitational field while magnetic torque tries to rotate an object to align with a magnetic field. For example, a compass needle rotates to align with the Earth's magnetic field (Halliday et al. (2014)). See Magnetic Torque and Gravitational Torque.

Torque Due to Gravity. For the purposes of MRI safety, the worst case gravitational torque magnitude is considered which is (Halliday et al. (2014), ASTM F2213-17 (2017))

$$\tau_g = ml \tag{T.1}$$

where τ_g is the magnitude of the torque due to gravity, m is the mass of the object, and l is the length of the object. If the magnetic torque on an object is less than the torque due to gravity, it is unlikely to pose a significant risk.

Torque Testing Methods. Standardized method for determining how much an object rotates to align with respect to the magnetic field of the scanner (ASTM F2213-17 (2017)).

(*MRSE*) Specified by ASTM publication F2213-17 (ASTM F2213-17 (2017)). Methods include Suspension, Low Friction Surface, Torsional Spring, Pulley, and Measured Displacement Force. The suspension and low friction methods cannot be used if the magnetic torque is greater than the gravitational torque. The test is conducted with each principle axis of the device oriented parallel to the bore. The test is conducted as near isocenter as possible.

Suspension: Suspend object from string and rotate. If no back rotation then no clinically significant torque.

Low Friction Surface: Place object on low friction surface outside magnet and measure coefficient of friction of surface by raising to the point where the object slides and measure the angle of incline θ. Use

$$u_f = \tan\theta \tag{T.2}$$

to determine the coefficient of friction u_f.

Place object with mass m on the surface in the center of the magnet and rotate. If no rotation then

$$\tau_m < l u_f m g \tag{T.3}$$

where τ_m is the magnitude of the magnetic torque, l is the object's length and g is the acceleration due to gravity.

Torsional Spring: Place object on holding platform supported by a torsional spring. Measure the angle outside the magnet. Place apparatus with object in the magnet, and measure the angle. Rotate the base and measure angles in 10 degree increments through 360 degrees. Determine the maximum torque from the angle closest to where the device swings from one equilibrium position to another using

$$\tau_m = k_s \Delta\theta \tag{T.4}$$

where k_s is the spring constant and $\Delta\theta$ is the change in angle from equilibrium.

Pulley: Place object on round rotating platform attached to pulley attached to force gauge. Measure with and without object. Read off force, and then calculate the torque τ_m using

$$\tau_m = r_p \left(F_T - F_f\right) \tag{T.5}$$

where r_p is the pulley radius, F_T is the maximum force measured with the device, and F_f is the frictional force of the platform without the device.

Displacement: Use the method for measuring the magnetic displacement force. See Displacement Force Testing. The object is suspended near the bore entrance at the bore, and the magnetic displacement force is calculated as described under Displacement Force Testing. An upper bound on the torque magnitude τ_{max} may then be estimated from

$$\tau_{max} = \frac{M_s F_m}{4\,|\nabla B_0|} \simeq \frac{2.2 F_m}{4\,|\nabla B_0|} \tag{T.6}$$

where M_s is the magnetization of a saturated object, F_m is the magnitude of the magnetic displacement force and $|\nabla B_0|$ is the magnitude of the maximum spatial gradient. $M_s \simeq 2.2$ for iron (ASTM F2213-17 (2017)).

Transdermal Drug Patches. Some drug therapies utilize patches or pads which deliver the drug transdermally. Some types contain metallic components which can potentially heat during MRI if in the rf volume, which can cause burns or alter the rate of drug delivery. If the patch is in the volume of excitation of the transmitting rf coil, the case should be reviewed with the responsible Level 2 MR Physician. Placement of an ice pack can be used to prevent heating, but as this may alter the rate of drug delivery, the decision to proceed this intervention should be made by the appropriate Level 2 MR Physician in consultation with the referring physician. Any decision to remove the patch must be accompanied by an order to do so from a physician caring for the patient (ACR Committee on MR Safety (2020)).

Transmetallation. When an ion is replaced by another ion in a chelate. For example, the gadolinium ion in a contrast molecule is replaced by another ion such as calcium or zinc leaving the free gadolinium ion able to bind to another ligand or receptor such as those for calcium (Ramalho et al. (2016), Rogowska et al. (2018)).

Transmission Line. An electrical conductor designed to carry alternating current (Rao (2009)). Some of the interactions of rf waves with implants or foreign bodies may be modeled as alternating current along a transmission line (Konings et al. (2000), Nyenhuis (2018)). See Antenna Effect.

Transmit Receive Coil. See Local Transmit Receive Coil.

Transmitter Bandwidth. Range of frequencies over which the rf excitation pulse acts (Brown et al. (2014)). The transmitter bandwidth depends upon the duration and shape of the radiofrequency pulse. To a first approximation the transmitter bandwidth $\Delta\omega_{rf} = 2\pi\Delta\nu_{rf}$ depends upon the pulse duration τ_p as

$$\Delta\omega_{rf} = \frac{2\pi}{\tau_p} \tag{T.7}$$

SAR increases with decreasing τ_p and hence increasing transmitter bandwidth (Shellock (2020)). See Specific Absorption Rate.

Traveling Wave. A wave which oscillates in time and travels in space. Electromagnetic radiation can be modeled as traveling waves (Rao (2009), Griffiths (2017)).

(*MRSE*) Traveling waves have forms such as

$$\mathbf{A}(z,t) = A_0 \cos(kz - \omega t)\widehat{\mathbf{x}} = Re\left[A_0 e^{i(kz-\omega t)}\right]\widehat{\mathbf{x}} \tag{T.8}$$

This indicates a wave with maximum amplitude A_0 which is traveling in the z direction with a wavelength $\lambda = 2\pi/k$ and a speed $v = \omega/k$ oscillating at a frequency ω directed along the x direction. Propagating electromagnetic waves can be modeled as traveling waves.

Treatment of Gadolinium Contrast Reactions. The ACR Manual on Contrast addresses treatment of contrast reactions due to gadolinium based contrast agents (ACR Committee on Drugs and Contrast Media (2020)). Treatment of acute allergic gadolinium reactions follows the same protocol as for iodinated agents. For mild reactions expectant management may be sufficient, and the reaction may abate on its own. The mainstays of pharmacological intervention are diphenhydramine and epinephrine (ACR Committee on Drugs and Contrast Media (2020)).

(*MRMD*) Some treatment strategies taken from the ACR Manual on Contrast and ACR Contrast Reaction Card include (ACR Committee on Drugs and Contrast Media (2020)):

Hives/diffuse erythema:

Monitor, vitals every 15 min, preserve IV access. If only skin findings but severe or progressive then 50 mg diphenhydramine PO/IM/IV (may worsen hypotension).

If associated hypotension or respiratory distress (anaphylaxis), give O2 6-10 L by face mask with 0.9% NS IV wide open and legs elevated, and give 0.3 cc 1:1000 epinephrine IM or 1 cc 1:10,000 epinephrine slowly IV flush or IV fluids. Call 911 or code blue.

Hypotension with tachycardia:

Vitals every 15 min, preserve IV access, O2 6-10 L by face mask with 0.9% NS IV wide open and legs elevated, and give 0.3 cc 1:1000 epinephrine IM or 1 cc 1:10,000 epinephrine slowly IV flush or IV fluids. Call 911 or code blue.

Laryngeal edema:

Monitor vitals, preserve IV access, O2 6-10 L by face mask, and give 0.3 cc 1:1000 epinephrine IM or 1 cc 1:10,000 epinephrine slowly IV flush or IV fluids. Call 911 or code blue.

Bronchospasm:

Monitor vitals, preserve IV access, O2 6-10 L by face mask, and give $\beta 2$ agonist inhaler (may repeat times 3). If not responsive to inhaler or severe give 0.3 cc 1:1000 epinephrine IM or 1 cc 1:10,000 epinephrine slowly IV flush or IV fluids. Call 911 or code blue.

Hypotension with bradycardia:

Monitor vitals, preserve IV access, O2 6-10 L by face mask with 0.9% NS IV wide open and legs elevated. If refractory give atropine 0.6 - 1.0 mg IV.

Treatment of Gadolinium Contrast Extravasation. Because of the chemical nature of gadolinium and small volumes injected, significant adverse sequelae from extravasation are extremely unlikely (ACR Committee on Drugs and Contrast Media (2020)).

Turbo Spin Echo(TSE). See Fast Spin Echo.

U

Unit Vector. (*MRSE*) A vector which has a magnitude of unity in some specified direction. Example: $r = (x, y, z) = (0, 0, 1)$. In Cartesian coordinates the unit vectors $(1, 0, 0)$, $(0, 1, 0)$, and $(0, 0, 1)$ are denoted $\widehat{\mathbf{x}}$, $\widehat{\mathbf{y}}$, and $\widehat{\mathbf{z}}$. In spherical coordinates, the unit vectors are denoted $\widehat{\boldsymbol{r}}$, $\widehat{\boldsymbol{\theta}}$, and $\widehat{\boldsymbol{\phi}}$ (Griffiths (2017)).

V

Vector. An object which is characterized by a magnitude and direction. Vectors are therefore used to describe properties such as force, velocity, acceleration, and magnetization (Halliday et al. (2014), Griffiths (2017))

Vector Field. (*MRSE*) A collection of vectors where at every point in space there is a property associated with a magnitude and direction. Vector fields are used to describe things which permeate space but vary from point to point in space, such as a magnetic field. Therefore, the magnetic field and the magnetic flux density are described by vector fields (Rao (2009)).

Ventriculoperitoneal Shunt. A device for regulating cerebrospinal fluid (CSF) pressure within the brain. Some models require radiography of the pressure dial to check for interval change in the pressure setting pre and post MRI as the force from the scanner can readjust the device pressure (Shellock (2020)).

Vertigo. Abnormal sensation of rotating motion. This has been reported by some individuals moving through the fields of MRI scanners. Postulated mechanisms include induced currents in the vestibular apparatus versus small torques in the inner ear receptors (Schenck and Kangarlu (2014)).

Volt (V). Unit for electric potential or electromotive force. $1\,V = 1\,J/C$ (Halliday et al. (2014)).

Volume Transmit Coil. A coil exciting a homogeneous volume over the volume of enclosed anatomy. See Body Transmit Coil.

W

Watt (W). Unit of power. $1\ W = 1\ J/s$ (Halliday et al. (2014)).

Wavelength (λ). In an electromagnetic wave the physical distance peak to peak of the wave. rf waves have relatively long wave lengths in vacuum and air on the order of hundreds of cm (Halliday et al. (2014)).

($MRSE$) In general

$$\lambda = v/\nu \tag{W.1}$$

where λ is wavelength, v is velocity, and ν is frequency.

In vacuum

$$\lambda_0 = c/\nu_0 \tag{W.2}$$

When traveling in a medium of dielectric constant ε_r the wavelength in the medium λ_m becomes (Dempsey et al. (2001), ASTM F2182-19e2 (2019))

$$\lambda_m = \frac{\lambda_0}{\sqrt{\varepsilon_r}} \tag{W.3}$$

The ε_r in a vacuum is 1, and the ε_r of tissue is around 80, so that the wavelength in tissue is on the order of 1/10 that in vacuum (Dempsey et al. (2001)).

Wave Number (k). ($MRSE$) Related to the inverse of thc wavelength, $k = 2\pi/\lambda$ (Rao (2009), Griffiths (2017)).

X

***x*-axis**. In some scanners with the patient supine head first in a horizontal field right-handed coordinate scanner the positive x-axis points toward the patient's right (Collins (2016)). In other scanners the positive x-axis points toward the patient's left hand (Brown et al. (2014), ASTM F2052-15 (2015)). Time varying gradients along the x-axis produce sagittal images. See Scanner Orientation.

Y

***y*-axis**. With the patient supine head first in a horizontal field scanner the positive *y*-axis points toward the ceiling of the scanner room (Brown et al. (2014), Collins (2016) ASTM F2052-15 (2015)). Time varying gradients along the *y*-axis produce coronal images. See Scanner Orientation.

Z

***z*-axis**. By convention this is the axis along the axis of the field which is along the bore of the magnet in a horizontal field system. In some scanners with the patient supine head first in a horizontal field scanner the positive z-axis points toward the head, and the positive $\mathbf{B}_0$ axis points toward the head (Collins (2016)). In other scanners the positive z-axis points along the feet (Brown et al. (2014), ASTM F2052-15 (2015)). Time varying gradients along the z-axis produce axial images. See Scanner Orientation.

Zone I. Area for unsupervised, unscreened individuals. Area safe for the general public (ACR Committee on MR Safety (2020)).

Zone II. Area for supervised, unscreened Non-MR personnel. This is the area where patients are typically screened and change into facility provided gowns or scrubs (ACR Committee on MR Safety (2020)).

Zone III. Area for supervised, screened individuals but not within the scanner room itself. This is the region where the presence of unscreened Non-MR personnel or objects could lead to serious injuries or death, and Zone III should be demarcated and clearly labelled as being potentially hazardous. Zone III must be physically restricted from general public access by a locking mechanism with the ability to differentiate between MR Personnel and Non-MR Personnel. Only MR Personnel should be given free access to Zone III, and there should be no exceptions to this guideline (ACR Committee on MR Safety (2020)).

Zone IV. The scanner room. Zone IV should be marked as hazardous due to the presence of a strong magnetic field. The door to the scanner room should be marked with a red light and sign lit at all times and on backup power with sign stating stating "The Magnet Is On". An exception to this would be a resistive magnet which loses field on loss of power to the magnet, but there should still be a red lit "The Magnet Is On" sign when the magnet is energized. The door to Zone IV should be closed except when it must be open for patient care or maintenance. When the door to Zone IV must remain open, a "caution" barrier for the door is recommended, such as straps or plastic chains across the doorway (ACR Committee on MR Safety (2020)).

Numerical

5 Gauss Exclusion Zone. Field strength corresponding to 0.5 mT (5 G) above which the FDA indicates there should be controlled access. Note that this does not necessarily coincide with the Zone concept in the ACR Manual on MR Safety (ACR Committee on MR Safety (2020)).

90-Degree Pulse. An rf pulse of sufficient duration to rotate the spin magnetization through a 90-degree angle. 90-degree pulses are used to initially excite magnetization in many pulse sequences (Brown et al. (2014)).

180-Degree Pulse. An rf pulse of sufficient duration to rotate the spin magnetization through a 180-degree angle. These are typically used to refocus magnetization or for initial excitation for inversion recovery sequences (Brown et al. (2014)).

Abbreviations

A. Ampere. Unit of current.

A. Scalar area.

A_0. Wave amplitude of traveling wave.

A_m. Fourier coeeficient.

ABMRS. American Board of Magnetic Resonance Safety.

ACR. American College of Radiology.

A_j. Surface area of object j.

AKI. Acute kidney injury.

ASRT. American Society of Radiologic Technologists.

ASTM. American Society for Testing and Materials International.

$\mathbf{A}$. Magnetic vector potential.

$\mathbf{A}(r,t)$. Magnetic vector potential as explicit function of space and time.

$\mathbf{A}(r)$. Spatial part of magnetic vector potential.

$\mathbf{A}(t)$. Wave as a function of time.

$\mathbf{A}_{Total_C}$. Standing wave with cosine dependence.

$\widetilde{A}_{Total_C}$. Phasor form of standing wave with cosine dependence.

$\mathbf{A}_{Total_S}$. Standing wave with sine dependence.

$\widetilde{A}_{Total_S}$. Phasor form of standing wave with sine dependence.

$\mathbf{A}(z,t)$. Arbitrary traveling wave.

a. Parameter in MDRD equation.

atm. Atmospheres.

B_m. Fourier coeeficient.

$\mathbf{B}$. Magnetic flux density.

$\mathbf{B}_0$. Magnetic flux density of the applied static field.

$\mathbf{B}_1$. Magnetic flux density of the applied rf field.

$\mathbf{B}_{1+}$. Magnetic flux density of the applied rf field in the same direction as precession.

$\mathbf{B}_{1-}$. Magnetic flux density of the applied rf field in the opposite direction as precession.

B_{1+rms}. Time averaged root mean square value of $\mathbf{B}_{1+}$.

$\mathbf{B}_{1ccw,lab}$. Applied rf field seen to rotate counter-clockwise by a laboratory observer.

$\mathbf{B}_{1cw,lab}$. Applied rf field seen to rotate clockwise by a laboratory observer.

$\mathbf{B}_{1cw,lab,Cir}$. Cicularly polarized applied rf field seen to rotate clockwise by a laboratory observer.

$\mathbf{B}_{1cw,lab,Lin}$. Linearly polarized applied rf field seen to rotate clockwise by a laboratory observer.

$\mathbf{B}_{1,lab}$. Applied rf field seen by an observer in the laboratory frame.

$\mathbf{B}_{1lab,Lin}$. Applied linear rf field seen by an observer in the laboratory frame.

B_{1x}. $\mathbf{B}_1$ amplitude factor in the x direction.

B_{1y}. $\mathbf{B}_1$ amplitude factor in the y direction.

$\widetilde{B}_{1x}$. $\mathbf{B}_1$ contribution to counter-rotating field in the x direction in phasor form.

$\widetilde{B}_{1y}$. $\mathbf{B}_1$ contribution to counter-rotating field in the y direction in phasor form.

$\mathcal{B}_{1x}$. $\mathbf{B}_1$ amplitude along x in phasor form.

$\mathcal{B}_{1y}$. $\mathbf{B}_1$ amplitude along y in phasor form.

B_A. Amplitude of magnetic flux density.

$\mathbf{B}_d(r, \theta)$. Magnetic dipole field.

$\mathbf{B}_h(z)$. Magnetic field along z-axis of Helmholz coil configuration.

$\mathbf{B}(\mathbf{r})$. Magnetic flux density expressed explicitly as a function of position vector $\mathbf{r}$.

$\mathbf{B}_s(z)$. Magnetic field along z-axis of two coil configuration.

BW. Bandwidth.

BW_{read}. Read bandwidth.

B_z. z component of magnetic flux density.

B_{zMax}. Maximum value of time varying gradient field.

$\mathbf{B}(z, t)$. Magnetic flux density as a function of z and t.

b. Parameter in the Pennes Bioheat Equation.

°C. Degrees Celsius. Unit of temperature.

C. Coulombs. Unit of charge. Also curve of integration. Also capacitance measured in Farads (F).

C_h. Chronaxie.

CIED. Cardiac implanted electronic device.

CKD. Chronic kidney disease.

CKD1-5. Stages of chronic kidney disease.

C_p. Heat capacity.

c. Speed of light.

cc. Cubic centimeters.

cgs. Old system of units.

cm. Centimeter. Unit of distance.

DWI. Diffusion weighted imaging.

$\mathbf{d}$. Dipole vector.

$d\mathbf{A}$. Differential area element.

dB. Decibel.

dB(A). A-weighted decibel.

dL. Deciliter.

$d\mathbf{l}$. Differential length element.

dV. Differential volume element.

$\mathbf{E}$. Electric field.

$\mathbf{E}_1$. Electric field of the applied rf field.

E_A. Amplitude of electric field.

$\mathbf{E}_d(r,\theta)$. Electric dipole field.

$\mathbf{E}_I$. Incident electric field.

$\mathbf{E}_{ind}$. Induced electric field.

EKG. Electrocardiogram.

EMI. Electromagnetic interference.

EPI. Echo planar imaging.

$\mathbf{E}_R$. Reflected electric field.

$\mathbf{E}(\mathbf{r})$. Electric field expressed explicitly as a function of position vector $\mathbf{r}$.

$\mathbf{E}_{Total}$. Total incident and reflected electric field.

ETL. Echo train length.

$\mathbf{E}(z,t)$. Electric field as a function of z and t.

$\mathcal{E}$. Energy.

$\mathcal{E}_i$. Energy deposited by rf pulse i.

$eGFR$. Estimated glomerular filtration rate.

emf. Electromotive force.

°F. degrees Fahrenheit.

F. Magnitude of force. Also Farad, unit of capacitance.

$\mathbf{F}$. Force.

FDA. Food and Drug Administration.

F_f. Magnitude of frictional force.

$\mathbf{F}_g$. Force due to gravity.

$\mathbf{F}_m$. Magnetic displacement force.

$\mathbf{F}_m(\mathbf{r})$. Magnetic displacement force as an explicit function of $\mathbf{r}$.

FOV. Field of view.

FOV_{fe}. Field of view in frequency encode direction.

FOV_{pe}. Field of view in phase encode direction.

F_p. Magnitude of the force product.

FSE. Fast spin echo.

FSEIR. Fast spin echo inversion recovery.

F_T. Measured force of pulley in pulley method of torque measurement.

f. Parameter in MDRD equation.

$f(r, \phi)$. Fourier series.

f_d. Duty cycle.

G. Gauss. Unit of magnetic flux density (cgs).

G_+. Magnitude of positive lobe of time varying gradient.

G_-. Magnitude of negative lobe of time varying gradient.

G_{fe}. Gradient in in frequency encode direction.

G_{Max}. Maximum gradient amplitude.

G_{Max_A}. Maximum gradient amplitude per axis.

G_{pe}. Gradient in in phase encode direction.

G_{ss}. Gradient in slice select direction.

G_{z_x}. Gradient of B_z with respect to x, dB_z/dx.

$G_{z_i Max}$. Maximum value of time varying gradient B_z with respcct to $i = x, y, z$.

GBCM. Gadolinium based contrast media.

GFR. Glomerular filtration rate.

GDD. Gadolinium deposition disease.

GRE. Gradient echo.

g. Gram. Unit of mass.

g. Acceleration due to gravity.

H. Henries. Unit of inductance ($kg\text{-}m^2/s^2\text{-}A^2$).

$\mathbf{H}$. Magnetic field.

$\mathbf{H}_0$. Applied static magnetic field.

Hz. Hertz. Unit of frequency.

$\mathbf{H}_d$. Demagnetizing field.

$\mathbf{H}_{in}$. Total field inside a body in an external field.

HRS. Heart Rhythm Society.

hr. Hour.

I. Ionic.

I. Current.

I_0. Current amplitude of a Hertzian dipole.

I_t. Current of a Hertzian dipole.

I_r. Rheobase.

I_{th}. Threshold generator current.

$\widetilde{I}(z)$. Current in phasor form.

$I(z, t)$. Traveling current wave.

IAC. Intersocietal Accreditation Commission.

ICD. Implantable Cardiac Defibrillator.

ICNIRP. International Commission on Nonionizing Radiation Protection.

IEC. International Electrotechnical Commission.

IM. Intramuscular.

IR. Inversion recovery.

ISO. International Organization for Standardization.

IV. Intravenous.

i. $\sqrt{-1}$

J. Joules. Unit of energy.

$\mathbf{J}$. Current density.

J_0. Current density amplitude.

$\mathbf{J}_s$. Surface currnt density.

JCAHO. Joint Commission on Accreditation of Healthcare Organizations.

K. Kelvin. Unit of temperature.

k. Wavenumber. $k = 2\pi/\lambda$.

kg. Kilograms. Unit of mass.

kHz. Kilohertz. Unit of frequency.

k_s. Spring constant of torsional spring.

kV. Kilovolt.

L. Liters. Also linear.

L. Inductance.

L01. Normal Operating Mode PNS stimulus limit.

L12. First Level Controlled Operating Mode PNS stimulus limit.

l. Length.

l_a. Length of axis of ellipsoid.

l_b. Length of axis of ellipsoid.

l_c. Length of axis of ellipsoid.

l_r. Resonant length.

M. Macrocyclic.

$\mathbf{M}$. Magnetization.

MAUDE. Manufacturer and User Facility Device Experience Database.

MDRD. Modified Diet in Renal Disease.

MHz. Megahertz. Unit of frequency.

MIPPA. Medicare Improvements for Patients and Providers Act of 2008.

MR. Magnetic resonance.

MRI. Magnetic resonance imaging.

MRMD. Magnetic resonance medical director.

MRSC™. Magnetic resonance safety certified.

MRSE. Magnetic resonance safety expert.

MRSO. Magnetic resonance safety officer.

M_S. Magnetization of a saturated object.

M_W. Molecular weight.

m. Mass. Also meters, unit of distance.

$\mathbf{m}$. Magnetic moment.

mg. Milligrams.

mL. Milliliter.

mm. Millimeters.

$\mathbf{m}_S$. Magnetic moment of a saturated object.

m_{S_z}. z-component of magnetic moment for a saturated object.

mT. Millitesla.

$\mathbf{m}_U$. Magnetic moment of an unsaturated object.

mW. Milliwatts.

N. Newton. Unit of force. Also nonionic.

N. Demagnetizing factor.

$\mathbf{N}$. Demagnetizing tensor.

$N_{a,b,c}$. Elements of demagnetizing tensor in principle axis system.

N_{acq}. Number of acquisitions.

N_E Number of echoes.

NEMA. National Electrical Manufacturers Association.

NEX. Number of excitations.

N_{ij}. General element of demagnetizing tensor.

NIOSH. National Institute for Occupational Safety and Health.

NRR. Noise reduction rating.

N_S. Number of slices.

NS. Normal saline.

NSF. Nephrogenic Systemic Fibrosis.

N_x. Number of points in the x direction.

N_y. Number of points in the y direction.

N_z. Number of points in the z direction.

n. Moles.

OSHA. Occupational Safety and Health Administration.

P. Pressure.

P. Power. Measured in Watts (W).

$\overline{P}$. Time averaged power over T_R.

P_0. Lowest sound pressure level detectable by human ear.

Pa. Pascal. Unit of pressure.

PNS. Peripheral nerve stimulation.

PO. Per os (by mouth).

P_t. Time averaged ohmic power.

$\mathbf{p}$. Electric dipole moment.

Q. Heat generated through metabolism.

Q_e. The charge enclosed.

Q_{e_j}. The charge enclosed for object j.

QRS. Label denoting parts of an EKG.

q. Charge.

$q_0(0)$. Point source charge at origin.

q_j. Charge on object j.

R. Euler rotation matrix.

R(θ). Euler rotation matrix as a function of $\beta = \theta$ with $\alpha = \gamma = 0$.

R. Resistance. Measured in Ohms (Ω). Also gas constant.

R_c. Radius of current loop.

R_E. Expansion ratio.

Re. Real part.

R_m. Mass ratio of exposed mass to body mass

R_s. Radius of sphere.

r. Radius.

rad. Radians.

rf. Radiofrequency.

rms. Root mean squared.

r_p. Pulley radius.

$\mathbf{r}$. Vector at observation point.

$\mathbf{r}'$. Vector at source point.

$\widehat{\boldsymbol{r}}$. Unit vector in spherical coordinates.

r_j. Radius of object j.

$\mathbf{r}_j$. Position of point charge at arbitrary location..

S. Siemens. Also surface of integration.

SAE. Specific Absorbed Energy.

SAR. Specific absorption rate.

$\overline{\text{SAR}}$. Time averaged specific absorption rate.

S_{Cr}. Serum creatinine.

SE. Spin echo.

SED. Specific energy dose.

SEL. Specific energy limit.

SNR. Signal-to noise ratio.

SR. Slew rate.

SR_{Max_A}. Maximum slew rate per axis.

$SR_{Max_{Eff}}$. Maximum effective slew rate.

s. Seconds. Unit of time.

s'. Location of Helmholtz coil.

T. Value of a specific time $t = T$.

T_a. Temperature in Pennes Bioheat Equation.

T_b. Blood temperature.

T. Tesla. Unit of magnetic flux density.

TE. Echo time.

TJC. The Joint Commission.

T_R. Repetition time.

T_T. Total time.

T/R. Transmit receive.

TSE. Turbo spin echo.

t. Variable time.

t_c. Tissue time constant.

t_i. Time period i during an image acquisition when there is no rf pulse.

t_{min}. Minimum time to transition between maximum and minimum gradient values.

t_r. Minimum rise time.

U. Magnetic potential energy.

u_f. Coefficient of friction.

V. Volts. Unit of electric potential. Also volume and volume of integration.

V_j. Electric potential of object j.

$V^{\pm}$. Voltage amplitudes traveling in opposite directions.

V_f. Flow potential.

V_G. Volume of cryogen gas.

V_L. Volume of cryogen liquid.

$V(\mathbf{r})$. Electric potential explicitly as a function of position vector $\mathbf{r}$.

$V(z, t)$. Traveling voltage wave.

$\widetilde{V}(z)$. Voltage in phasor form.

v. Velocity scalar.

$\mathbf{v}$. Velocity vector.

W. Watts. Unit of power.

WB. Whole body.

$\widehat{\mathbf{x}}$. Unit vector along x in the laboratory frame.

$\widehat{\mathbf{x}}'$. Unit vector along x in the rotating frame.

$\widehat{\mathbf{y}}$. Unit vector along y in the laboratory frame.

Z. Impedance.

Z_0. Intrinsic impedence.

z'. Location on the z-axis of a coil.

$\widehat{\mathbf{z}}$. Unit vector along z in the laboratory frame.

α. Angle. May refer to phase. Euler angle.

α_p. Pulse flip angle.

β. Phase. Also Euler angle.

γ. Gyromagnetic ratio. Also Euler angle.

Δx. Slice thickness in the x direction.

Δy. Slice thickness in the y direction.

Δz. Slice thickness in the z direction.

$\Delta\omega$. Bandwidth.

$\Delta\omega_{fe}$. Frequency encode bandwidth.

$\Delta\omega_{pe}$. Phase encode bandwidth.

$\Delta\omega_{rf}$. Transmitter bandwidth.

$\Delta\omega_{fe}$. Bandwidth in the frequency encode direction.

$\Delta\omega_{pe}$. Bandwidth in the phase encode direction.

Ψ. Magnetic scalar potential.

ε. Electric permittivity.

ε_0. Electric permittivity of free space.

ε_r. Relative electric permittivity also known as the dielectric constant.

ζ. Ratio of principle axis demagnetizing factors for ellipsoid.

θ. Angle.

$\widehat{\boldsymbol{\theta}}$. Unit vector in spherical coordinates. θ is the angle between the z-axis and r.

κ. Thermal conductivity.

λ. Wavelength.

λ_0. Wavelength in vacuum.

λ_m. Wavelength in a given medium.

μ. Magnetic permeability.

μ_0. Magnetic permeability of free space.

μ_f. Coefficient of friction.

μ_r. Relative magnetic permeability.

μs. Microseconds.

ν. Frequency in Hz.

ν_0. Frequency in vacuum.

ν_m. Frequency in a given medium.

ρ. Mass density.

ρ_c. Resistivity.

ρ_j. Electric charge density.

ρ_q. Volume electric charge density.

σ_c. Conductivity.

σ_q. Surface charge density.

σ_{q_j}. Surface charge density for charge q on object j.

τ. Stimulus time in strength duration curve.

τ_g. Magnitude of torque due to gravity.

τ_{p_i}. Pulse time of the ith pulse in a pulse train.

τ_m. Magnitude of magnetic torque.

τ_{max}. Magnitude of maximum magnetic torque.

$\boldsymbol{\tau}_m(\mathbf{r})$. Magnetic torque as an explicit function of $\mathbf{r}$.

$\boldsymbol{\tau}_m$. Magnetic torque.

τ_p. Pulse time.

$\widehat{\boldsymbol{\phi}}$. Unit vector in spherical coordinates. ϕ lies in the xy-plane.

χ_E. Scalar electric susceptibility.

χ_m. Scalar magnetic susceptibility.

χ_E. Electric susceptibility.

$\boldsymbol{\chi}$. Magnetic susceptibility tensor.

$\boldsymbol{\chi}_e$. Magnetic susceptibility tensor in the laboratory frame with respect to $\mathbf{H}_0$.

$\boldsymbol{\chi}_{ij}$. General element of demagnetizing tensor.

$\boldsymbol{\chi}_m$. Magnetic susceptibility tensor with respect to $\mathbf{H}_{in}$.

$\boldsymbol{\chi}_p$. Magnetic susceptibility tensor in the principle axis system.

ω. Angular frequency in radians/second.

ω_0. Angular precession frequency in radians/second.

ω_1. Angular frequency of applied rf field in radians/second.

Ω. Ohms. Unit of resistance or impedance.

Bibliography

ABMRS. The Organization, https://abmrs.org/about-abmrs/the-organization. accessed april 27, 2020.

ACR (2015). *ACR Magnetic Resonance Imaging Quality Control Manual.* American College of Radiology.

ACR (2018). *MRI Accreditation Program Requirements.* American College of Radiology.

ACR Committee on Drugs and Contrast Media (2020). *ACR Manual On Contrast Media.* American College of Radiology.

ACR Committee on MR Safety (2020). *ACR Manual On MR Safety, version 1.0.* American College of Radiology.

Allison, J. and Yanasak, N. (2015). What MRI sequences produce the highest specific absorption rate (SAR), and is there something we should be doing to reduce the SAR during standard examinations? *Am. J. Roentgenol.*, 205:W140–W140.

Ansorge, R. and Graves, M. J. (2016). *The Physics and Mathematics of MRI.* Morgan & Claypool Publishers.

Arfken, G. B. and Weber, H. J. (2005). *Mathematical Methods for Physicists.* Elsevier.

ASTM F2052-15 (2015). *Standard Test Method for Measurement of Magnetically Induced Displacement Force on Medical Devices in the Magnetic Resonance Environment.* American Society for Testing and Materials (ASTM) International.

ASTM F2119-07 (2013). *Standard Test Method for Evaluation of MR Image Artifacts from Passive Implants.* American Society for Testing and Materials (ASTM) International.

ASTM F2182-19e2 (2019). *Standard Test Method for Measurement of Radio Frequency Induced Heating on or Near Passive Implants During Magnetic Resonance Imaging.* American Society for Testing and Materials (ASTM) International.

ASTM F2213-17 (2017). *Standard Test Method for Magnetically Induced Torque on Medical devices in the Magnetic Resonance Environment.* American Society for Testing and Materials (ASTM) International.

ASTM F2503-20 (2020). *Standard Practice for Marking Medical Devices and Other Items for Safety in the Magnetic Resonance Environment.* American Society for Testing and Materials (ASTM) International.

Baddoo, N. (2008). Stainless steel in construction: A review of research, applications, challenges and opportunities. *J. Constr. Steel Res.*, 64:1199–1206.

Barrett, M. and Macdonald, A. (1999). The form

of magnetic work in a fundamental thermodynamic equation for a paramagnet. *Am. J. Phys.*, 67:613–615.

Becerra, R. J., Harsh, M. J., Boskamp, E. B., Guclu, C. C., Duby, T., Havens, T. J., Hunks, R. S., P, N. J., Peterson, W. T., Piepenburg, S. J., Stormont, R. S., and Vavrek, R. M. (2006). Intrumentation: Magnet, gradients, and coils. In Edelman, R. R., Hesselink, J. R., Zlatkin, M. B., and Crues III, J. V., editors, *Clinical Magnetic Resonance Imaging*. Saunders Elsevier.

Bennett, M. C., Wiant, D. B., Gersh, J. A., Dolesh, W., Ding, X., Best, R., and Bourland, J. (2012). Mechanisms and prevention of thermal injury from gamma radiosurgery headframes during 3T MR imaging. *J. Appl. Clin. Med. Phys.*, 13:54–70.

Boyer, T. H. (1988). The force on a magnetic dipole. *Am. J. Phys.*, 56:688–692.

Brown, R. W., Haacke, E. M., Cheng, Y.-C. N., Thompson, M. R., and Venkatesan, R. (2014). *Magnetic Resonance Imaging: Physical Principles and Sequence Design.* John Wiley & Sons, 2cd edition.

Calamante, F., Ittermann, B., Kanal, E., Norris, D., and on MR Safety, I. W. G. (2016). Recommended responsibilities for management of MR safety. *J. Magn. Reson. Imaging*, 44:1067–1069.

Carluccio, G., Bruno, M., and Collins, C. M. (2016). Predicting long-term temperature increase for time-dependent SAR levels with a single short-term temperature response. *Magn. Reson. Med.*, 75:2195–2203.

Chakeres, D. W., Kangarlu, A., Boudoulas, H., and Young, D. C. (2003). Effect of static magnetic field exposure of up to 8 Tesla on sequential human vital sign measurements. *J. Magn. Reson. Imaging*, 18:346–352.

Clausen, L., Cook, J., Garlock, A., Perkins, A., Pierce, B., Seitz, K., Turk, J., Worthington, A., Lipman, E., and Odle, T. (2018). Radiologic technologist best practices for MR safety https://www.asrt.org/docs/default-source/research/whitepapers/asrt18_mrsafety-whitepaper.pdf. Accessed April 30, 2019.

Coey, J. M. (2010). *Magnetism and Magnetic Materials.* Cambridge University Press.

Collins, C. M. (2016). *Electromagnetics in Magnetic Resonance Imaging: Physical Principles, Related Applications, and Ongoing Developments.* Morgan & Claypool Publishers.

Collins, C. M. and Smith, M. B. (2001). Calculations of B1 distribution, SNR, and SAR for a surface coil adjacent to an anatomically-accurate human body model. *Magn. Reson. Med.*, 45:692–699.

Cullity, B. D. and Graham, C. D. (2011). *Introduction to Magnetic Materials.* John Wiley & Sons.

Delfino, J. G., Krainak, D. M., Flesher, S. A., and Miller, D. L. (2019). MRI-related FDA adverse event reports: a 10-yr review. *Medical Physics*, 46(12):5562–5571.

Dempsey, M. F. and Condon, B. (2001). Thermal injuries associated with MRI. *Clin. Radiol.*, 56:457–465.

Dempsey, M. F., Condon, B., and Hadley, D. M. (2001). Investigation of the factors responsible for burns during MRI. *J. Magn. Reson. Imaging*, 13:627–631.

Dewangan, A., Patel, A., Bhadania, A., et al. (2015). Stainless steel for dairy and food industry. *RJAHDS*, 6:66–71.

Faulkner, W. H. J. (2016). *Rad Tech's Guide to MRI: Basic Physics, Instrumentation, and Quality Control.* John Wiley & Sons.

FDA (2014). *Criteria for significant risk investigations of magnetic resonance diagnostic devices.* US Food and Drug Administration.

Fernow, R. C. (2016). *Principles of Magnetostatics.* Cambridge University Press.

Golestanirad, L., Angelone, L. M., Kirsch, J., Downs, S., Keil, B., Bonmassar, G., and Wald, L. L. (2019). Reducing rf-induced heating near implanted leads through high-dielectric capacitive bleeding of current (CBLOC). *IEEE Transactions on Microwave Theory and Techniques.*

Greenberg, T. D., Hoff, M. N., Gilk, T. B., Jackson, E. F., Kanal, E., McKinney, A. M., Och, J. G., Pedrosa, I., Rampulla, T. L., et al. (2019). Acr guidance document on mr safe practices: Updates and critical information 2019. *Journal of Magnetic Resonance Imaging*, 51(2):331–338.

Gregory, T. S., Cheng, R., Tang, G., Mao, L., and

Tse, Z. T. H. (2016). The Magnetohydrodynamic Effect and Its Associated Material Designs for Biomedical Applications: A State-of-the-Art Review. *Adv. Funct. Mater.*, 26:3942–3952.

Griffiths, D. J. (2017). *Introduction to Electrodynamics*. Cambridge University Press.

Halliday, D., Resnick, R., and Walker, J. (2014). *Fundamentals of Physics*. John Wiley & Sons.

Hardy, K. (2015). Training staff in MRI safety. *Radiology Today*, 16:20.

Hayt, W. H. and Buck, J. A. (2012). *Engineering Electromagnetics*. McGraw-Hill New York, 8th edition.

Horwood, L., Attili, A., Luba, F., Ibrahim, E.-S. H., Parmar, H., Stojanovska, J., Gadoth-Goodman, S., Fette, C., Oral, H., and Bogun, F. (2017). Magnetic resonance imaging in patients with cardiac implanted electronic devices: focus on contraindications to magnetic resonance imaging protocols. *EP Europace*, 19(5):812–817.

Hoult, D. (2000). The principle of reciprocity in signal strength calculations—a mathematical guide. *Concepts Magn. Reson.*, 12:173–187.

Ilott, A. J. and Jerschow, A. (2019). Aspects of NMR reciprocity and applications in highly conductive media. *Concepts Magn. Reson. Part A*, 47A:e21466.

Indik, J. H., Gimbel, J. R., Abe, H., Alkmim-Teixeira, R., Birgersdotter-Green, U., Clarke, G. D., Dickfeld, T.-M. L., Froelich, J. W., Grant, J., Hayes, D. L., et al. (2017). 2017 HRS Expert

Consensus Statement on Magnetic Resonance Imaging and Radiation Exposure in Patients with Cardiovascular Implantable Electronic Devices. *Heart Rhythm*, 14:97–153.

International Commission on Non-Ionizing Radiation Protection (1998). ICNIRP guidelines for limiting exposure to time-varying electric, magnetic, and electromagnetic fields (up to 300 GHz). *Health Phys.*, 74:494–522.

International Commission on Non-Ionizing Radiation Protection (2004). Medical magnetic resonance (MR) procedures: protection of patients. *Health Phys.*, 87:197–216.

International Commission on Non-Ionizing Radiation Protection (2009a). Amendment to the ICNIRP "Statement on medical magnetic resonance (MR) procedures: protection of patients". *Health Phys.*, 97:259–261.

International Commission on Non-Ionizing Radiation Protection (2009b). Guidelines on limits of exposure to static magnetic fields. *Health Phys.*, 96:504–514.

International Commission on Non-Ionizing Radiation Protection (2009c). ICNIRP statement on the "guidelines for limiting exposure to time-varying electric, magnetic, and electromagnetic fields (up to 300 GHz)". *Health Phys.*, 97:257–258.

International Commission on Non-Ionizing Radiation Protection (2010). Guidelines for limiting exposure to time-varying electric and magnetic fields (1 Hz to 100 kHz). *Health Phys.*, 99(6):818–836.

International Commission on Non-Ionizing Radiation Protection (2013). ICNIRP guidelines on limits of exposure to laser radiation of wavelengths between 180 nm and 1,000 μm. *Health Phys.*, 105:271–295.

International Commission on Non-Ionizing Radiation Protection (2014). ICNIRP guidelines for limiting exposure to electric fields induced by movement of the human body in a static magnetic field and by time-varying magnetic fields below 1 Hz. *Health Phys.*, 106:418–425.

International Commission on Non-Ionizing Radiation Protection (2017). ICNIRP statement on diagnostic devices using non-ionizing radiation: existing regulations and potential health risks. *Health Phys.*, 112:305–321.

International Commission on Non-Ionizing Radiation Protection (2020). Guidelines for limiting exposure to electromagnetic fields (100 kHz to 300 GHz). *Health Physics*, 118(5):483–524.

International Electrotechnical Commission (2010). International standard, medical equipment—IEC 60601-2-33: particular requirements for the safety of magnetic resonance equipment for medical diagnosis.

International Electrotechnical Commission (2013). Electroacoustics—IEC 61672-1: sound level meters- part 1- specifications.

International Electrotechnical Commission (2014).

Safety of laser products- part 1—IEC 60825-1: equipment classification and requirements.

International Electrotechnical Commission (2015). International standard, amendment 2, medical equipment—IEC 60601-2-33: particular requirements for the safety of magnetic resonance equipment for medical diagnosis.

International Organization for Standardization (2013). *ISO226:2013 Acoustics- normal equal-loudness contours.*

International Organization for Standardization (2018). *ISO TS 10974 Assessment of the safety of magnetic resonance imaging for patients with an active implantable medical device.*

Jiles, D. (2015). *Introduction to Magnetism and Magnetic Materials.* CRC press.

Jin, J. (1999). *Electromagnetic Analysis and Design in Magnetic Resonance Imaging*, volume 1. CRC press.

Junk, R. (2014). MRI facility design to support MRI safety. In Shellock, F. G. and Crues III, J. V., editors, *MRI Bioeffects, Safety, and Patient Management*, chapter 7, pages 171–194. Biomedical Research Publishing Group.

Kainz, W. (2014). MRI standards and guidance documents from the United States. In Shellock, F. G. and Crues III, J. V., editors, *MRI Bioeffects, Safety, and Patient Management.* Biomedical Research Publishing Group.

Kanal, E. and Shaibani, A. (1994). Firearm safety

in the MR imaging environment. *Radiology*, 193:875–876.

Keene, M. N. (2014). Using ferromagnetic detection systems in the MRI environment. In Shellock, F. G. and Crues III, J. V., editors, *MRI Bioeffects, Safety, and Patient Management*, chapter 13, pages 299–327. Biomedical Research Publishing Group.

Keevil, S. (2014). MRI standards and safety guidelines in Europe. In Shellock, F. G. and Crues III, J. V., editors, *MRI Bioeffects, Safety, and Patient Management*, chapter 29, pages 665–677. Biomedical Research Publishing Group.

Knoepfel, H. E. (2000). *Magnetic Fields: A Comprehensive Theoretical Treatise for Practical Use.* John Wiley & Sons.

Konings, M. K., Bartels, L. W., Smits, H. F., and Bakker, C. J. (2000). Heating around intravascular guidewires by resonating rf waves. *J. Magn. Reson. Imaging*, 12:79–85.

Langman, D. A., Goldberg, I. B., Finn, J. P., and Ennis, D. B. (2011). Pacemaker lead tip heating in abandoned and pacemaker-attached leads at 1.5 tesla MRI. *J. Magn. Reson. Imaging*, 33:426–431.

Liang, Z.-P. and Lauterbur, P. C. (2000). *Principles of Magnetic Resonance Imaging: A Signal Processing Perspective.* IEEE Press.

Lide, D. R., editor (2004). *CRC Handbook of Chemistry and Physics*, volume 85. CRC press.

Lo, K. H., Shek, C. H., and Lai, J. (2009). Recent

developments in stainless steels. *Mater. Sci. Eng. R. Rep.*, 65:39–104.

Mandel, N. S., Ramdial, J. L., and Marcus, E. N. (2017). A second-degree burn after MRI. *Clev. Clin. J. Med.*, 84:348.

Mattei, E., Gentili, G., Censi, F., Triventi, M., and Calcagnini, G. (2015). Impact of capped and uncapped abandoned leads on the heating of an MR-conditional pacemaker implant. *Magn. Reson. Med.*, 73:390–400.

Mattei, E., Triventi, M., Calcagnini, G., Censi, F., Kainz, W., Mendoza, G., Bassen, H. I., and Bartolini, P. (2008). Complexity of MRI induced heating on metallic leads: experimental measurements of 374 configurations. *Biomed. Eng. Online*, 7:11.

MAUDE (2020). https://www.accessdata.fda.gov/scripts/cdrh/cfdocs/cfmaude.

McDonald, R. J., Levine, D., Weinreb, J., Kanal, E., Davenport, M. S., Ellis, J. H., Jacobs, P. M., Lenkinski, R. E., Maravilla, K. R., Prince, M. R., et al. (2018). Gadolinium retention: a research roadmap from the 2018 NIH/ACR/RSNA workshop on gadolinium chelates. *Radiology*, 289:517–534.

McJury, M. (2014). Acoustic noise and MRI procedures. In Shellock, F. G. and Crues III, J. V., editors, *MRI Bioeffects, Safety, and Patient Management*, chapter 4, pages 88–130. Biomedical Research Publishing Group.

McRobbie, D. (2012). Occupational Exposure in MRI. *Br. J. Radiol.*, 85:293–312.

McRobbie, D. (2014). Bioeffects of gradient magnetic fields. In Shellock, F. G. and Crues III, J. V., editors, *MRI Bioeffects, Safety, and Patient Management*, chapter 3, pages 63–87. Biomedical Research Publishing Group.

McRobbie, D. W. (2020). *Essentials of MRI Safety.* John Wiley & Sons.

McRobbie, D. W., Moore, E. A., Graves, M. J., and Prince, M. R. (2017). *MRI from Picture to Proton.* Cambridge University Press.

MIPPA (2008). Medicare Improvements for Patients and Providers Act of 2008 Pub. L. No. 110-275.

Mueller, L. J. (2011). Tensors and rotations in NMR. *Concepts Magn. Reson. Part A*, 38A:221–235.

Muthalaly, R. G., Nerlekar, N., Ge, Y., Kwong, R. Y., and Nasis, A. (2018). MRI in patients with cardiac implantable electronic devices. *Radiology*, 289:281–292.

NIOSH, (2019). NIOSH hierarchy of controls. https://www.cdc.gov/niosh/topics/hierarchy.

Nyenhuis, J. A. (2014). Using MRI simulations and measurements to evaluate active implants. In Shellock, F. G. and Crues III, J. V., editors, *MRI Bioeffects, Safety, and Patient Management*, chapter 21, pages 485–513. Biomedical Research Publishing Group.

Nyenhuis, J. A. (2018). Electromagnetic distribution in tissue with conductive devices. In Shrivastava, D., editor, *Theory and Applications of Heat Transfer In Humans*, chapter 12, pages 203–231. John Wiley & Sons.

Occupational Safety and Health Administration 29 CFR Standard 1910.145. *Specifications for accident prevention signs and tags.* Occupational Safety and Health Administration.

Occupational Safety and Health Administration Standard 29 CFR 1910.301. *Electrical.* Occupational Safety and Health Administration.

Occupational Safety and Health Administration Standard 29 CFR 1910.95. *Occupational noise exposure.* Occupational Safety and Health Administration.

Occupational Safety and Health Administration Standard 29 CFR 1910.97. *Nonionizing radiation.* Occupational Safety and Health Administration.

Occupational Safety and Health Administration Standard 29 CFR 1915.12. *Permit required confined spaces.* Occupational Safety and Health Administration.

Padmanabhan, D., Kella, D. K., Mehta, R., Kapa, S., Deshmukh, A., Mulpuru, S., Jaffe, A. S., Felmlee, J. P., Jondal, M. L., Dalzell, C. M., et al. (2018). Safety of magnetic resonance imaging in patients with legacy pacemakers and defibrillators and abandoned leads. *Heart Rhythm*, 15(2):228–233.

Panych, L. P. and Madore, B. (2018). The physics of MRI safety. *J. Magn. Reson. Imaging*, 47:28–43.

Parsh, J., Seth, M., Aronow, H., Dixon, S., Heung, M., Mehran, R., and Gurm, H. S. (2015). Choice of estimated glomerular filtration rate equation impacts drug-dosing recommendations and risk stratification in patients with chronic kidney disease undergoing percutaneous coronary interventions. *J. Am. Coll. Cardiol.*, 65:2714–2723.

Pictet, J., Meuli, R., Wicky, S., and van der Klink, J. J. (2002). Radiofrequency heating effects around resonant lengths of wire in MRI. *Physics in Medicine and Biology*, 47(16):2973–2985.

Ramalho, J., Semelka, R., Ramalho, M., Nunes, R., AlObaidy, M., and Castillo, M. (2016). Gadolinium-based contrast agent accumulation and toxicity: an update. *Am. J. Neuroradiol.*, 37:1192–1198.

Rao, N. N. (2009). *Fundamentals of Electromagnetics for Electrical and Computer Engineering.* Pearson Prentice Hall.

Reinking, D. (2014). Performing sedation and anesthesia during MRI. In Shellock, F. G. and Crues III, J. V., editors, *MRI Bioeffects, Safety, and Patient Management*, chapter 15, pages 351–367. Biomedical Research Publishing Group.

Riley, K. F., Hobson, M. P., and Bence, S. J. (2006). *Mathematical Methods for Physics and Engineering: A Comprehensive Guide.* Cambridge University Press.

Rogowska, J., Olkowska, E., Ratajczyk, W., and Wolska, L. (2018). Gadolinium as a new emerging contaminant of aquatic environments. *Environ. Toxicol.*, 37:1523–1534.

Sackett, C. A. (2018). *Statistical Mechanics and Thermodynamics: A Practical Approach*. Zero K Press.

Saraswat, A. K. and Smith, M. A. (2014). MRI screening for patients and individuals. In Shellock, F. G. and Crues III, J. V., editors, *MRI Bioeffects, Safety, and Patient Management*, chapter 12, pages 282–298. Biomedical Research Publishing Group.

Schaefer, D. (2014). Bioeffects of radiofrequency power deposition. In Shellock, F. G. and Crues III, J. V., editors, *MRI Bioeffects, Safety, and Patient Management*, chapter 5, pages 131–154. Biomedical Research Publishing Group.

Schenck, J. F. (2000). Safety of strong, static magnetic fields. *J. Magn. Reson. Imaging*, 12:2–19.

Schenck, J. F. and Kangarlu, A. (2014). Bioeffects of static magnetic fields. In Shellock, F. G. and Crues III, J. V., editors, *MRI Bioeffects, Safety, and Patient Management*, chapter 2, pages 29–63. Biomedical Research Publishing Group.

Semelka, R. C., Ramalho, J., Vakharia, A., AlObaidy, M., Burke, L. M., Jay, M., and Ramalho, M. (2016). Gadolinium deposition disease: initial description of a disease that has been around for a while. *Magn. Reson. Imaging*, 34:1383–1390.

Shellock, F. G. (2020). *Reference Manual for Magnetic Resonance Safety, Implants, and Devices: 2020 Edition.* Biomedical Research Publishing Group.

Shellock, F. G., Kanal, E., and Gilk, T. B. (2011). Regarding the value reported for the term "spatial gradient magnetic field" and how this information is applied to labeling of medical implants and devices. *Am. J. Roentgenol.*, 196:142–145.

Shrivastava, D. (2018). A generic thermal model for perfused tissues. In Shrivastava, D., editor, *Theory and Applications of Heat Transfer In Humans*, chapter 1, pages 3–13. John Wiley & Sons.

Simon, J., Amde, M., and Poggio, E. D. (2011). Interpreting the estimated glomerular filtration rate in primary care: benefits and pitfalls. *Cleve. Clin. J. Med.*, 78:189–195.

Spaeth, R. B. and Gollub, R. L. (2014). Claustrophobia, anxiety, and emotional distress in the MRI environment. In Shellock, F. G. and Crues III, J. V., editors, *MRI Bioeffects, Safety, and Patient Management*, chapter 8, pages 195–216. Biomedical Research Publishing Group.

Steer, M. (2019). *Microwave and RF Design.* NC State University.

Suprihanto, A. (2017). Magnetic properties of austenitic stainless steel 316l and 316lvm after high temperature gas nitriding treatment. *ROTASI*, 19:72–75.

Tsai, L. L., Grant, A. K., Mortele, K. J., Kung, J. W., and Smith, M. P. (2015). A practical guide to MR imaging safety: what radiologists need to know. *Radiographics*, 35:1722–1737.

Wasserman, A. L. (2011). *Thermal Physics: Concepts and Practice.* Cambridge University Press.

Zangwill, A. (2013). *Modern Electrodynamics.* Cambridge University Press.

Zélis, P. M., Vega, V., Prida, V., Costa-Arzuza, L., Béron, F., Pirota, K., López-Ruiz, R., and Sánchez, F. (2017). Effective demagnetizing tensors in arrays of magnetic nanopillars. *Phys. Rev. B*, 96:174–427.

Made in the USA
Las Vegas, NV
03 November 2026

57421606R00186